Springer Theses

Recognizing Outstanding Ph.D. Research

Aims and Scope

The series "Springer Theses" brings together a selection of the very best Ph.D. theses from around the world and across the physical sciences. Nominated and endorsed by two recognized specialists, each published volume has been selected for its scientific excellence and the high impact of its contents for the pertinent field of research. For greater accessibility to non-specialists, the published versions include an extended introduction, as well as a foreword by the student's supervisor explaining the special relevance of the work for the field. As a whole, the series will provide a valuable resource both for newcomers to the research fields described, and for other scientists seeking detailed background information on special questions. Finally, it provides an accredited documentation of the valuable contributions made by today's younger generation of scientists.

Theses are accepted into the series by invited nomination only and must fulfill all of the following criteria

- They must be written in good English.
- The topic should fall within the confines of Chemistry, Physics, Earth Sciences, Engineering and related interdisciplinary fields such as Materials, Nanoscience, Chemical Engineering, Complex Systems and Biophysics.
- The work reported in the thesis must represent a significant scientific advance.
- If the thesis includes previously published material, permission to reproduce this must be gained from the respective copyright holder.
- They must have been examined and passed during the 12 months prior to nomination.
- Each thesis should include a foreword by the supervisor outlining the significance of its content.
- The theses should have a clearly defined structure including an introduction accessible to scientists not expert in that particular field.

More information about this series at http://www.springer.com/series/8790

Supervisor's Foreword

Power and particle exhaust are major challenges on the way to a nuclear fusion reactor based on magnetic confinement of high-temperature plasmas. The removal of the helium ash of the DT fusion process can be facilitated by guiding the plasma to dedicated plasma-facing components (so-called divertors), resulting in high particle flux densities to the plasma-facing surfaces and the build-up of high neutral pressures, needed to pump helium neutrals. With the concentration of the particle flux to the first wall of the fusion reactor, enormous power flux densities come along, altogether putting severe demands on the first wall materials.

Tungsten is the main candidate for the plasma-facing material in the divertor because of its high thermal conductivity, high melting point, small uptake of fuel atoms and low erosion rate, the last of which determines the lifetime of the first wall components, thus the availability of a future reactor. For tungsten, erosion is determined by physical sputtering under normal operating conditions, setting-in if the energy of impinging ions is larger than the sputtering threshold, which depends on the mass ratio of projectiles and wall target atoms. To optimize the lifetime of the first wall, the plasma temperature in front of the components has to be low enough that the energy of the impinging ions does not surpass the energy threshold for sputtering.

Under such operating conditions, fluctuations of the plasma temperature may dominate the overall erosion rate of tungsten, in particular via intermittent transport events which have been observed in the plasma boundary of fusion devices but also in linear plasma devices. As a consequence, an experimental characterization of the excursions of the plasma temperature due to such transport events (called "blobs") is required to assess their importance for tungsten sputtering. The lifetime of such events on the order of tens to hundreds of microseconds brings about a large complexity for such measurements.

In his dissertation, Dr. Hubeny determined the electron temperature within fluctuating structures in the linear plasma device PSI-2 by Thomson scattering and applied for the first time a technique called conditional averaging to Thomson scattering signals as the electron temperature within blobs cannot be determined during one single event because of the relatively moderate electron density in PSI-2.

Using this technique, he could quantify temperature fluctuations and intermittent transport events. The results show that the fluctuation level and the size of intermittent transport events could well give rise to significant material erosion even if the average plasma temperature corresponds to ion energies below the sputtering threshold. These findings may influence the choice of operational conditions in future fusion reactors based on magnetic confinement.

Jülich, Germany
November 2018

Prof. Dr. Bernhard Unterberg

Abstract

For the viable operation of a thermonuclear fusion reactor, large pressure gradients are necessary in the edge of a magnetically confined plasma device. Turbulent cross-field transport across these gradients severely limits the durability of the current reactor designs by the erosion of plasma-facing components. The convective character of this transport is caused by intermittently occurring plasma filaments, which are observed with similar characteristics in the edge of a wide range of plasma devices. Besides the ballistic plasma transport, filaments lead to pressure fluctuation statistics with a positive skewness and kurtosis, i.e. short, local spikes in density and temperature time traces. Erosion predictions based on time-averaged measurements underestimate dynamic effects on erosion mechanisms, e.g. the sputtering yield, due to its nonlinear temperature dependence.

The aim of this thesis was the development of a time-resolved laser Thomson scattering diagnostic system to investigate plasma dynamics leading to and during intermittent transport events, which were identified at the linear plasma generator PSI-2 in high-powered, steady-state deuterium discharges with fast visual imaging. To retain the temporal information in the Thomson scattering set-up, the signal was recorded in a triple grating spectrometer on a shot-to-shot basis with a photon counting method. The synchronization of the Nd:YAG laser and spectrometer to a fast framing camera allowed the novel usage of conditional averaging to create Thomson spectra. The selection of a subset of laser pulses with conditions in time and space was based on the plasma state characterized by 100 captured frames around each laser pulse with a time resolution of 3 μs.

Time-averaged Thomson scattering profiles of plasma density and temperature were obtained for deuterium, helium, neon and argon discharges at different high-power settings, and a reasonable agreement was found with results from the Langmuir probe as the standard diagnostic tool at PSI-2. Stand-alone fast camera measurements were used to characterize the dynamics in the visible wavelength range and found distinct mode structures and slower oscillations with increasing ion mass. Two synchronized fast cameras showed that these modes are elongated along the magnetic field.

The application of conditional averaging on Thomson scattering was successfully tested argon discharges, where rotating brightness fluctuations with a 75 μs. correlation time were found to correspond to a 20% temperature fluctuation amplitude around the temporal mean at the profile maxima. Subsequently, filaments in the edge of high-power, low gas-feed deuterium discharges were selected by conditional averaging and a significant temperature increase was found in the edge of Thomson scattering profiles upon ejection, accompanied by a 10% drop in bulk density.

Finally, the fluctuation statistics from the fast camera were found comparable to the corresponding density and temperature traces obtained by the time-resolved Thomson scattering. A range of positively skewed temperature distributions was constructed and used to calculate the influence on the sputtering yield.

List of Publications

Papers

S. Elgriw, **M. Hubeny**, A. Hirose and C. Xiao, *Effect of resonant magnetic perturbations on edge plasma parameters in the STOR-M tokamak*, Radiat. Eff. Defects S., in press (2018)
M. Rack, D. Höschen, D. Reiter, B. Unterberg, J. W. Coenen, S. Brezinsek, O. Neubauer, S. Bozhenkov, G. Czymek, Y. Liang, **M. Hubeny**, Ch. Linsmeier, and the Wendelstein 7-X Team, *Probe manipulators for Wendelstein 7-X and their interaction with the magnetic topology*, Plasma Sci. Technol., Vol. 20, No. 5 (2018)
M. Hubeny, B. Schweer, D. Lugenhölscher, U. Czarnetzki, B. Unterberg, *Thomson scattering of plasma turbulence on PSI-2*, Nucl. Mater. Energy, Vol. 12, pp. 1253–1258 (2017)
X. Jiang, G. Sergienko, N. Gierse, **M. Hubeny**, A. Kreter, S. Brezinsek, Ch. Linsmeier, *Design and development of a LIBS system on linear plasma device PSI-2 for in situ real-time diagnostics of plasma-facing materials*, Nucl. Mater. Energy, Vol. 12, pp. 1224–1230 (2017)
N. Gierse, M. Z. Tokar, S. Brezinsek, T. F. Giesen, **M. Hubeny**, A. Huber, V. Philipps, A. Pospieszczyk, G. Sergienko, J. Wegner, Q. Xiao, U. Samm, Ch. Linsmeier and the TEXTOR Team *Time resolved imaging of laser induced ablation spectroscopy (LIAS) in TEXTOR and comparison with modeling* Physica Scripta, Vol. 167, 014034 (2016)
A. Terra, G. Sergienko, **M. Hubeny**, A. Huber, Ph. Mertens, V. Philipps and the TEXTOR Team, *High heat-flux self-rotating plasma-facing component: Concept and loading test in TEXTOR*, Journal of Nuclear Materials, Vol. 463, pp. 1252–1255 (2015)
O. Mitarai, Y. Ding, **M. Hubeny**, et al., *Plasma current sustainment after iron core saturation in the STOR-M tokamak*, Fusion Eng. Des., Vol. 89, pp. 2467–2471 (2014)
M. Dreval, **M. Hubeny**, et al., *Plasma confinement modification and convective transport suppression in the scrape-off layer using additional gas puffing in the STOR-M tokamak*, Plasma Phys. Contr. F., Vol. 55, 035004 (2013)
S. Elgriw, C. Xiao, **M. Hubeny**, et al., *Studies of Resonant Magnetic Perturbations in the STOR-M Tokamak*, 39 th EPS Conference (2012)

Talks

M. Hubeny et al., *Turbulence evaluation at PSI-2 by fast visible imaging*, DPG **2014**, Berlin
M. Hubeny et al., *Turbulence evaluation at PSI-2 by fast visible imaging*, TEC-Meeting **2014**, Ghent
M. Hubeny et al., *Turbulence evaluation at PSI-2*, DPG **2015**, Bochum

M. Hubeny et al., *Thomson Scattering of Plasma Turbulence at PSI-2*, PMIF **2015**, Jülich
M. Hubeny et al., *Thomson Scattering of Plasma Turbulence on PSI-2*, Gaseous Electronic Conference (GEC) **2016**, Bochum

Posters

M. Hubeny et al., *Thomson Scattering of Plasma Turbulence on PSI-2*, Plasma-Surface-Interactions (PSI) **2016**, Rome
M. Hubeny et al., *Time-resolved Thomson Scattering of Plasma Filaments in PSI-2*, Plasma-Facing Materials and Components (PFMC) **2017**, Neuss

Acknowledgements

I would like to express my deepest gratitude to everybody who supported, motivated and inspired me throughout my Ph.D. time. Meiner Familie danke ich besonders für Ansporn in den richtigen Momenten und für die Nachsicht, wenn die Kommunikation wieder mal etwas zu dürftig ausfiel.

A big thanks go to my Supervisor and Tutor Bernhard Unterberg for this challenging topic and for his trust in a successful completion. I want to thank Bernd Schweer for his help with words and deeds and for significantly easing the interaction with the workshop. I am very grateful for the excellent spectrometer provided by my External Supervisors Uwe Czarnetzki and Dirk Luggenhölscher. I want to thank all of you for the extensive help and support in understanding many detailed physics issues and questions around the complex topic of the Ph.D. project. For the theoretical side of the physical understanding, I want to especially thank Dirk Reiser for his help and collaboration.

My office inmates Bruno Jasper and Mitja Beckers deserve special thanks for the good atmosphere in the office and in our doctoral shared flat. Furthermore, I would like to thank Tobias Wegener and Sören Möller for the intense physics discussions and continuous stream of crazy ideas. Thank you, Timo Dittmar and HR Koslowski, for providing excellent Linux support in our small institute community. Michael Vogel, Sebastian Kraus and Arkadi Kreter deserve much appreciation for the operation and maintenance of the PSI-2. Furthermore, I want to thank Michael Reinhart, Luxherta Buzi, Maren Hellwig, Rahul Rayaprolu, Simon Heuer, as well as all other colleagues of the IEK-4 for the nice working environment.

Last but not least, I would like to especially thank my life partner Kristin Kirschke for her loving support during the whole duration of the Ph.D. time.

Contents

Chapter 1
Introduction

Fusion

A major milestone to be achieved in our current civilization is the controlled usage of nuclear fusion as a prime source of energy. In the course of history numerous high cultures initially expanded and prospered, but eventually collapsed as they were agriculturally bound and thus too susceptible to the prevailing climates [1]. Only in the current anthropocene we are tapping into a one-time cheap and plenty energy source compressed over millennia in the form of fossil fuels. Unprecedented technological advancements during the industrial revolution swept away the agricultural restrictions and permitted humanity to grow even beyond global sustainability in terms of most resources. Moreover, the ever rising global energy demand will inevitably drain the more ecological-friendly fossil fuels, which enables the earth hosting (currently) 10 (7) bn humans, who are incidentally as skilled and educated as never before. This historic “sweet spot” empowers us already to harvest various renewable resources, yet acknowledging western per-capita energy, resource and technology consumption to all humanity will likely lead to a total exhaustion of fossil fuels with disastrous ramifications. Consequently, steering towards an emission-free, safe and abundant energy production is of utmost importance.

As a power base-load and hence avoiding over-excessive energy storage capabilities [2, 3] nuclear fusion should be developed for promising the lowest ecological impact after overcoming the immense technological challenge of building a fusion power plant. Furthermore, independence of solar powered/weather-dependent renewable energy sources will become a substantial reassurance once the climate change is in full swing with more erratic and extreme weather conditions.

Nuclear fusion converts the binding energy of nucleons in atomic cores to kinetic energy by changing the potential energy of the nucleons involved in the fusion process. Fusing light nuclei leads to the highest energy conversion per nucleon (in contrast to fission) with an energy density about six orders of magnitude larger compared to the energy density stored in chemical bonds. This energy density and the short interaction range of the strong force allow our sun to release energy for billions of

M. Hubeny, *The Dynamics of Electrons in Linear Plasma Devices and Its Impact on Plasma Surface Interaction*, Springer Theses,
https://doi.org/10.1007/978-3-030-12536-3_1

years. Serendipitously, the most abundant fusion fuel in the universe, hydrogen, has a dramatically ($\sim 10^{20}$ [4]) lower reaction probability (cross-section) compared to its isotopes, Deuterium and Tritium (DT), which extremely slows down the hydrogen fusion process in our sun. DT fusion on the other hand has, by a two magnitudes margin, the highest reaction cross-section amongst all fusion reactions involving light species [5] and is therefore the reaction of choice for a fusion reactor on earth:

$$\mathrm{D} + \mathrm{T} \rightarrow {}^{4}\mathrm{He} + \mathrm{n} + 17.6\,\mathrm{MeV}. \tag{1.1}$$

A resonant transition to the intermediate ^{5}He compound nucleus [6, Chap. 2] causes an enhanced reaction cross-section peak at a projectile energy of roughly 80 keV. Although the temperature needed for fusion reactions to occur is even lower at roughly 20 keV due to the high energy tail in a Maxwell-Boltzmann distribution, achieving this temperature requires the fuel to be held in the plasma state with unconventional and exceptional insulation and heat shields. The high kinetic energies and temperatures encountered in the field of plasma physics led to the common usage of electron volt as unit of temperature or energy with $\mathrm{T[eV]} = \frac{1}{k_\mathrm{B}}\,\mathrm{T[K]} \cong 11600\,\mathrm{T[K]}$, where the Boltzmann constant k_B relates the average kinetic energy to a temperature.

The abundance of Deuterium is exuberant as the natural isotope ratio to hydrogen is 0.0156% and thus sea water contains a virtually endless supply. Tritium on the other hand undergoes radioactive decay with a half-life of 12.7 years, but it is produced in small quantities in fission reactors and is planned to be bred from lithium in the vicinity of the fusion reactor by energetic neutrons via

$${}^{6}\mathrm{Li} + \mathrm{n} \rightarrow {}^{4}\mathrm{He} + \mathrm{T} + 4.8\,\mathrm{MeV}, \tag{1.2}$$

$${}^{7}\mathrm{Li} + \mathrm{n} \rightarrow {}^{4}\mathrm{He} + \mathrm{T} + \mathrm{n} - 2.5\,\mathrm{MeV}. \tag{1.3}$$

These neutrons are readily available from the fusion reaction in Eq. 1.1 and carry four-fifths of the total released energy due to momentum conservation, while the α-particles (^{4}He) carry the remaining one-fifth and are designated to maintain the fuel temperature.

The high abundance of the Deuterium and lithium and the absence of radioactive products in the reaction itself makes controlled nuclear fusion an important ("nuclear") solution for the energy crisis of the 21st century [7]. To gain access to fusion as a means of electricity production a power plant ought to create an environment in which a self-sustained fusion reaction is generated, fueled and the released energy captured. The particle velocities needed for the above reactions to take place require the fuel to be in a fully ionized, hot plasma state [8]. The reaction rate determined by density and temperature of the plasma must be high enough to heat the plasma by thermalization of the α-particles, while the neutrons are extracted and absorbed outside the reaction chamber to create heat for a conventional steam cycle and breed Tritium. There are a number of reactor approaches reaching towards reactor grade fusion rates without the massive gravitational force in a burning star. From extremely short and repetitive-explosive to steady state operated, either transient confinement by inertia (e.g. in bombs/explosions) or (electro-)magnetic confinement are under active investigation, of which the latter is related to this work.

Plasma Confinement

The core difficulty in nuclear fusion is reaching and maintaining the temperatures needed to overcome the strong coulomb repulsion between the like-charged ions, Deuterium and Tritium. The required ion temperatures of 10 to 20 keV need plasma containment without material contact and good heat insulation. By restricting the motion of the ionized particles with magnetic fields through the Lorentz force the heat conductivity perpendicular to the applied field is strongly reduced and thereby the fusing hot center can be insulated from the solid walls of a reactor vessel [9]. Minimizing the losses of heat and particles leads to the important figure of energy and particle confinement times, describing how long it takes for heat and particles to diffuse or convect across the magnetic field, whereas the parallel dynamics of the plasma are unchanged by the magnetic field in first-order approximation.

While there are reactor concepts with open magnetic field line configurations, the most promising designs, the tokamak and stellarator, close the field lines in a toroidal loop, forming closed magnetic flux surfaces. Since the surfaces are nested around the central loop, ideally no particles escape via their unrestricted motion parallel along the magnetic field. Both concepts are shown next to each other in Fig. 1.1, where the stellarator immediately draws attention by its complicated structure. Generally, a toroidal and poloidal magnetic field is necessary for a basic plasma confinement. The tokamak requires a toroidal current to create the poloidal magnetic field, which is beneficial for initial ohmic (resistive) heating, but makes this concept prone to instabilities connected to this mega-ampere current. The stellarator's three dimensional coils create this poloidal field without plasma current and allow an unrestricted steady state operation, while the tokamak needs additional non-inductive current drive. The obvious downside of the stellarator is the challenging engineering and the lesser

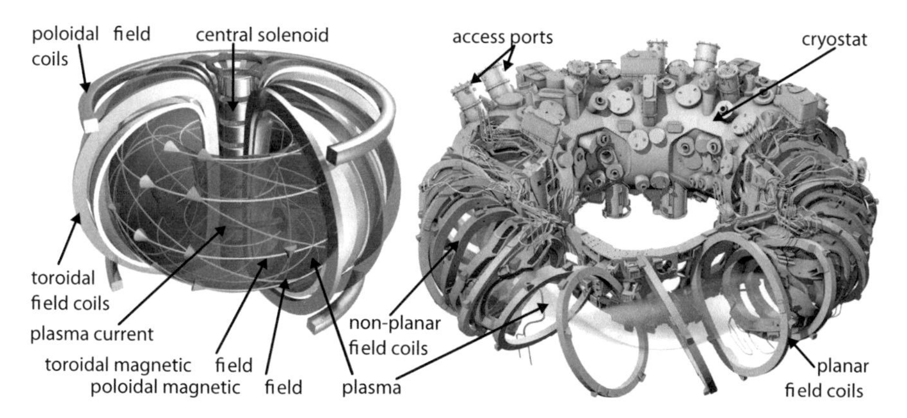

Fig. 1.1 Schematic of tokamak (left) and stellarator (right) with basic components (with courtesy of C. Brandt and C. Bickel, MPG IPP)

number of (large) experiments put the concept about 1.5 generations behind the tokamak.

Fusion Power Plant

Technically, the popular "50 years to fusion" have actually been achieved already in the 1990s when the tokamaks TFTR and JET (worlds largest) produced several MW of fusion power [10, 11], proving that fusion with a magnetic confinement is feasible [12]. The stellarator LHD lacked permission to use Tritium as fuel but reached similar plasma conditions, achieving a Q-factor, the ratio between fusion and heating power, of $Q = 0.2$, while JET conditions were closer to unity (break-even) for a shorter time ($\sim$1 s) at $Q = 0.6$. Consequently, the 50 years following these initial fusion achievements are now devoted to eliminate all remaining obstacles preventing a robust and commercially viable fusion power plant [13].

A burning DT-plasma uses the α-particles to sustain its temperature and the energetic neutrons absorbed in the blanket to power a conventional steam cycle, while the neutrons must be utilized to create Tritium from lithium (Eqs. 1.2 and 1.3). For the operation of a fusion plant the burning chamber must be constantly fueled with DT and the created α-particle must be pumped out after equilibrating (heating) with the plasma, since it would otherwise dilute the plasma. The technology to create all necessary parts of the first burning fusion reactor ITER is readily available, partially because established concepts were used for the superconducting coils, plasma heating and structural components. The immense fluxes of energetic neutrons (14 MeV) are unprecedented and thus blanket modules are tested in ITER as one of the final milestones. The single most crucial and still unresolved issue of fusion is the plasma transport and all it entails, from fusion performance and confinement of α-particles in the core to the plasma edge and its interaction with the solid wall.

Plasma Wall Interaction

The scientific progress over the last decades led to revised reactor geometries [15], several techniques to control the stability of plasma discharges and a number of not fully understood regimes of reduced transport, called transport barriers [16]. The basic magnetic and device geometry used by most current and likely future tokamaks is depicted as a poloidal cross-section in Fig. 1.2, recognizable by the x-point above the divertor region. This shape creates a defined point of maximum heat and particle load, whereas the main chamber requirements are reduced accordingly. The edge transport barriers act at the last closed flux surface (LCFS) to ensure a low pressure in the scrape-off-layer (SOL), where the plasma is transported down to the outer leg of the divertor and pumped out. This primary exhaust point is subject to intense research, as the engineering constraints on the maximum power load and additional cooling capabilities strongly influence the whole reactor design and operation.

Turbulence and Enhanced Erosion

Turbulence increases particle and energy transport across the magnetic field lines vastly and it appears on the macroscopic down to the microscopic level. Transport barriers reduce turbulent transport significantly by limiting the structure size to the

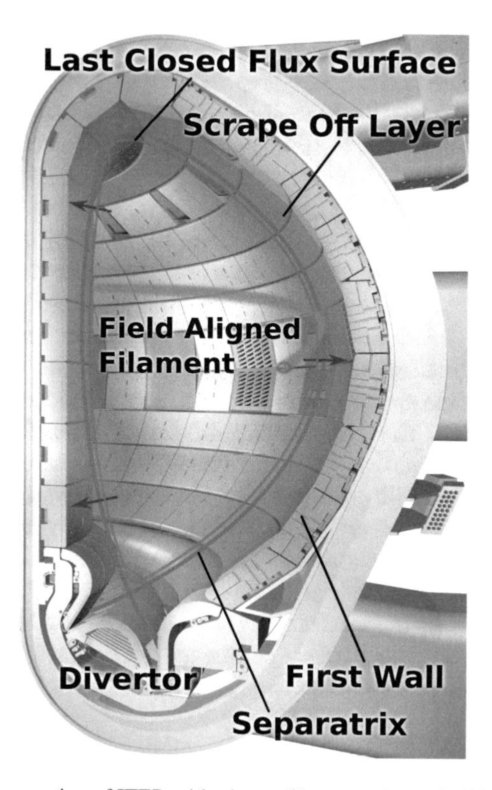

Fig. 1.2 Poloidal cross-section of ITER with plasma filament schematic [14]

scale on which they act, though leaving turbulent meso- and microscopic structures intact [17]. Historically, these structures have been called blobs, avaloids or filaments, of which the last is now the most adapted term due to their elongation along the magnetic field and rather compact radial and poloidal size as sketched in Fig. 1.2. While they have been found in virtually all magnetically confined devices, the SOL plasma is increasingly comprised of these structures in experiments with high performance discharges [18]. Since this filamentary transport is rather convective in its nature it poses a fundamental problem to the first wall and the divertor. A steady and homogeneous flux of plasma into the SOL and divertor can be modeled and used as the input for erosion predictions, but how does the erosion and wall recycling change when accounting for turbulent filamentary transport? Since the physical processes involved in wall erosion are extremely sensitive to plasma parameters, e.g. physical sputtering [19], the exact dynamics of the plasma transport are essential for a valid erosion prediction and must be understood in detail.

Tungsten is the wall material of choice for its high melting temperature, low erosion, moderate Tritium retention and absence of stable chemical compounds with Deuterium or Tritium [20]. Usually, the design heat flux reaching the divertor is based on its maximum heat load capacity, while the flux towards the first wall is supposed to be limited by the fast parallel transport [21]. Hence the plasma temperatures reached at the first wall should be below the sputtering threshold due to the mass ratio between tungsten and Deuterium or Tritium. However, filaments can carry hot plasma from the LCFS towards the wall due to their high velocity and surpass the temperature threshold. Furthermore, a SOL entirely constituted by filaments is more permeable for charge-exchange neutrals, which are another important erosion channel, while the large surface of the first wall holds a large potential for dynamic release of impurities.

Plasma filaments are observed as intermittent bursts of higher plasma pressure in the edge of tokamaks and stellarator, but they are also present in devices with purely toroidal magnetic field and linear plasma generators. While several described mechanisms for filament generation exist depending on plasma parameters (gradients) and magnetic configuration [22], the similarity in the statistical properties leads to employment of the same diagnostics to investigate the filaments, independent of the magnetic configuration. The limited availability time of reactor sized experiments restricts their use for testing long-term erosion effects, while linear plasma devices are steady state operated with pressures reaching the conditions at the main chamber first wall and divertor [23]. Investigating filaments in linear machines could thus clarify their potential effects on erosion as well as testing theories for turbulent transport in a somewhat simpler and better defined plasma environment. Therefore, physical properties and the effects of filaments on plasma wall interactions can be studied in detail at linear machines, where the experimental access and plasma conditions are more flexible and cost-effective compared to tokamaks or stellarator.

Aim of the Work

Most of the currently existing diagnostics suitable for filament measurements are limited to giving qualitative conclusions. Thomson scattering is a quantitative and non-intrusive method, but usually limited to time-averaged measurements in low density plasmas, yet considered one of the most accurate plasma density, temperature and velocity measurements. The aim of this thesis is the characterization of electron dynamics in the edge region of the linear plasma device PSI-2 by establishing a proof-of-principle for (pseudo) time-resolved measurements of intermittent transport events with Thomson scattering for the first time. Turbulent dynamics important for plasma wall interaction processes are introduced in Chap. 2 with a comparison of state-of-the-art diagnostics at various magnetically confined plasma devices. The light scattering theory is derived in Chap. 3, while the general experimental setup at PSI-2 and the newly installed Thomson scattering setup are described in Chaps. 4 and 5.

A photon counting method is developed for Thomson scattering retaining temporal information for the combination with the imaging, by which the pseudo time-resolved measurement is realized by conditionally averaging over the intermittent plasma dynamics. These signal processing and basic calibration methods are explained in Chap. 6.

Discharge conditions suitable for creating intermittent transport are sought after by fast optical imaging and subsequently characterized. The laser aided diagnostic system for Thomson Scattering is newly installed at PSI-2 and the measurements are compared to existing diagnostics in Chap. 7.

The conditional average Thomson scattering method is applied to argon and Deuterium discharges. The specific and characterized dynamics analyzed by Thomson scattering are complemented with the statistical measures by the fast optical imaging to evaluate the impact on erosion by intermittent plasma transport in Chap. 8. The key question behind the thesis is whether the investigated micro-turbulence is a significant contributor to erosion provided that the physical parameters governing erosion processes are comparable in fusion reactor and linear plasma generators.

References

1. Tainter J (1990) The collapse of complex societies. Cambridge University Press, Cambridge
2. Wagner F (2014) Electricity by intermittent sources: an analysis based on the german situation 2012. Eur Phys J Plus 129(20):20
3. Romanelli F (2016) Strategies for the integration of intermittent renewable energy sources in the electrical system. Eur Phys J Plus 131:53. (article id. 53, 15 pp)
4. Adelberger EG, Austin SM, Bahcall JN et al (1998) Solar fusion cross sections. Rev Mod Phys 70(4):1265–1291
5. Bosch H-S, Hale GM (1992) Improved formulas for fusion cross-sections and thermal reactivities. Nucl Fusion 32(4):611–631
6. Kikuchi M (2011) Frontiers in fusion research: physics and fusion. Springer, London
7. Chen FF (2011) An indispensable truth. An indispensable truth: how fusion power can save the planet. Springer Science+Business Media, LLC, n/a 2011. ISBN 978-1-4419-7819-6
8. Piel A (2010) Plasma physics. Plasma physics: an introduction to laboratory, space, and fusion plasmas. Springer, Berlin, n/a 2010. ISBN 978-3-642-10490-9
9. Naujok D (2006) PMI in controlled fusion. Atomic, optical, and plasma physics, vol 39. Springer, Berlin
10. Strachan JD, Batha S, Beer M et al (1997) TFTR DT experiments. Plasma Phys Control Fusion 39(12B):B103–B114
11. Rebut P-H (1992) The JET preliminary tritium experiment. Plasma Phys Control Fusion 34(13):1749–1758
12. Ongena J, Koch R, Wolf R, Zohm H (2016) Magnetic-confinement fusion. Nat Phys 12:398–410
13. Federici G, Kemp R, Ward D et al (2014) Overview of eu demo design and r&d activities. Fusion Eng Des 89(7):882–889. (Proceedings of the 11th international symposium on fusion nuclear technology-11 (ISFNT-11) Barcelona, Spain, 15–20 September 2013)
14. ITER Organization, Poloidal cut of ITER vacuum vessel. http://www.iter.org/
15. Kadomtsev BB (1988) Evolution of the tokamak. Plasma Phys Control Fusion 30(14):2031–2049
16. Wagner F (2007) A quarter-century of H-mode studies. Plasma Phys Control Fusion 49(12B):B1–B33
17. Tynan GR, Fujisawa A, McKee G (2009) Topical review: a review of experimental drift turbulence studies. Plasma Phys Control Fusion 51(11):113001. (article id. 113001, 77 pp)
18. D'Ippolito DA, Myra JR, Zweben SJ (2011) Convective transport by intermittent blob-filaments: comparison of theory and experiment. Phys Plasmas 18(6):060501. (article id. 060501, 48 pp)
19. Behrisch R (2007) Sputtering by particle bombardment : experiments and computer calculations from threshold to MeV energies, vol 110. Topics in applied physics. Springer, Berlin
20. Brezinsek S, Loarer T, Philipps V et al (2013) Fuel retention studies with the ITER-like wall in JET. Nucl Fusion 53(8):083023

21. Pitts RA, Carpentier S, Escourbiac F et al (2011) Physics basis and design of the ITER plasma-facing components. J Nucl Mater 415(1);Supplement:957. (Proceedings of the 19th international conference on plasma-surface interactions in controlled fusion)
22. Krasheninnikov SI, D'Ippolito DA, Myra JR (2008) Recent theoretical progress in understanding coherent structures in edge and SOL turbulence. J Plasma Phys 74(5):679–717
23. Kreter A (2011) Reactor-relevant plasma-material interaction studies in linear plasma devices, Fusion Sci Technol 59:51–56. (8th International conference on open magnetic systems for plasma confinement (OS2010), Novosibirsk, Russia, 05–09 July 2010)

Chapter 2
Plasma Wall Transition Dynamics

The transition from the hot plasma edge to the solid first wall and divertor region is particularly complex for its extreme parameter gradients involving the fusion fuel, ash and impurity species cooling down the edge plasma and recycling in the plasma wall transition region [1]. Since the gradients are dynamically changing over several orders of magnitudes in time and space, capturing the prevailing physical processes is challenging, especially when non-local effects require large computation volumes, yet fine spatiotemporal resolution. Turbulence acting on multiple scales limits the applicability of linear models to modeling equilibria, while predictions and physical understanding need the inclusion of non-linear effects.

Turbulent transport as one of the main obstacles in fusion research is dominating the edge and SOL region of magnetically confined devices, where a myriad of instabilities can be excited in different devices depending on multiple device parameters [2]. Consequently, statistical models and scaling laws are trying to unravel and describe the dynamics to make limited predictions. The tremendous effort to understand these processes in the edge and SOL is twofold: In the edge pedestal the pressure rises 2–3 orders of magnitudes over a few cm, representing a "standing wave" of energy capable of melting kilograms of wall material on collapse and thus severely impair the reactor operation, complicated by the fact that the pedestal stability is caused by turbulence and self-organized fluctuations itself [3]. Secondly, turbulent structures passed on from the pedestal towards the wall make the SOL cross-field conductive, opposing its function to shield the first wall by transporting plasma down to the divertor.

Since plasma material interaction and turbulent transport are vast fields of research only the most important PWI processes are explained in the following. The influence of turbulence on these processes motivates the effort for the advanced diagnostic setup presented in this thesis. The drift wave instability will serve as an example for turbulence generation. Additionally, the turbulent structures and current plasma edge diagnostics are compared regarding similarities between toroidal reactor and linear plasma devices, since the influence of turbulence on the PWI processes needs to be reconciled between a real reactor and the simplified surrogate environment.

M. Hubeny, *The Dynamics of Electrons in Linear Plasma Devices and Its Impact on Plasma Surface Interaction*, Springer Theses,
https://doi.org/10.1007/978-3-030-12536-3_2

2.1 Plasma Wall Interactions

The lifetime of the first wall and divertor is a major issue for the duty cycle of a whole reactor power plant. Erosion of the first wall and divertor is ultimately limiting the plasma operation, hence the processes leading to erosion are of prime interest [4]. While neutrons are mostly passing through the first wall, plasma particles, neutrals and radiation are absorbed or reflected. For particle collisions the exact interaction process depends strongly on the projectile velocity and the involved species, defining the momentum transfer. Tungsten is the preferred material for most of the plasma facing components for its high melting temperature and its atomic mass, which is highly favorable for two-body collisions with light species. A major downside is the radiative power of tungsten as a core impurity due to its many electronic levels as high-Z material, which are excited once it is transported into the hot core plasma.

Figure 2.1 shows a sketch of chemical and physical sputtering for fuel, ash and impurities. The absence of chemical reactions with the DT-fuel is another advantage of tungsten over the previously favored but very reactive carbon, which acquires an unacceptably high amount of Tritium (bound on the surface as co-deposits and retained in the bulk material), since Tritium is radioactive and one of the scarce goods in a fusion reactor. The tungsten surface can be strongly modified by heat and particle loads, leading to blisters, dendrites or cracks [5, 6]. These structures are then less tightly connected to the surface and accommodate fuel and impurities through the increased effective surface area. Therefore, plasma impact can release these adsorbed species at much lower projectile energies compared to physical sputtering. The release of neutral or ions depends on the plasma dynamics, which then further dictate whether these particles are merely adsorbed without much impact or return to the surface with higher energy leading to net erosion.

Tungsten has an average atomic mass of 183.84 amu and based on two body collisions it repels light projectiles like Deuterium up to a certain threshold energy, which is defined by the surface/bulk binding energy and the mass ratio. When surpassing this threshold energy, the number of sputtered atoms per incoming ion, the sputtering yield, increases dramatically over several magnitudes. At the high energy limit, ions are increasingly implanted into the bulk material, adding to the so-called

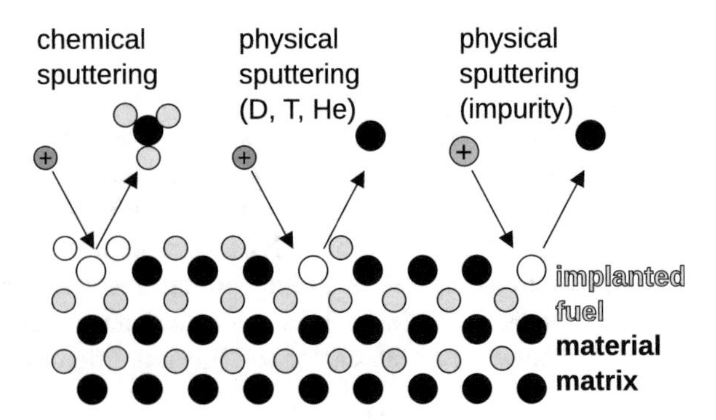

Fig. 2.1 Sputtering processes at plasma wall transition (with courtesy of S. Brezinsek, FZJ)

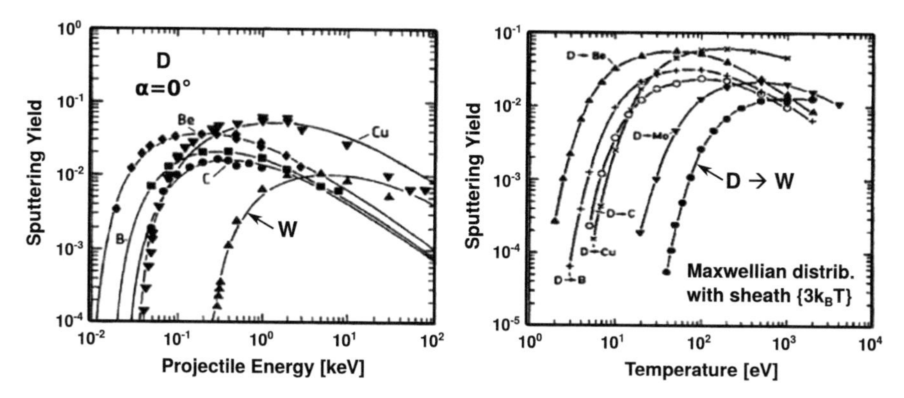

Fig. 2.2 Sputtering yield versus kinetic energy and temperature of Deuterium impinging on tungsten and other surfaces [7]

(fuel) retention in the wall, while the sputtering yield decreases again. This behavior can be seen for all species with corresponding threshold energies and is depicted in Fig. 2.2, where the sputtering yield is shown for specific projectile energies of different species impinging on tungsten on the left and for a Maxwellian energy distribution of Deuterium on various surface materials on the right. SOL temperatures in the sub-threshold region up to 20 to 30 eV promise negligible sputtering rates, which is therefore used as a boundary condition for reactor operation.

The sputtering rates resolved for projectile energies and temperatures are readily available for a vast number of material combinations [7]. While impurities account for only a small fraction of the plasma, their share in sputtering and thus total erosion can be significant, since their mass is five (Helium-Neon) to twenty (Deuterium-Argon) times higher. The trajectories of charged particles are governed by the magnetic field and an often used oblique angle towards the surface increases the chance of reflection and minimizes the momentum transfer.

However, the generation of neutrals can occur via charge exchange (CX) in the edge/SOL, leading to energetic neutral particles, which are not affected by the magnetic field and hit the surface directly along their path if no collisions change the momentum. The further inside (towards the LCFS) the neutral is created the more energetic it is statistically, especially in the pedestal region of the edge. On average the net effect on erosion is determined by the penetration depth, based on species densities in the edge/SOL and the species involved in the CX process. While heavier impurities would cause tremendous damage for being energetic and neutral, their higher charge state makes this process unlikely to happen.

The basic inputs for PWI models like ERO [8], calculating the erosion of first wall and divertor materials, are steady state profiles of plasma pressure and flux in the SOL. Impurity fractions have to be estimated and the CX process is determined by the recycling of plasma at the wall, while the penetration depth towards the pedestal is again estimated based on the pressure profiles. The rates of PWI processes are thus calculated by the time-average pressure and flux profiles. Large scale transient effects have been considered in case of plasma disruptions and found that their tremendous heat and particle load must be mitigated. Small scale fluctuations are not considered so far. Key questions for modeling the PWI are the retention of fuel in and the erosion of the first wall material. Especially the retention of tritium is critical as there is a maximum permitted amount for a given reactor for safety reasons.

Leaving aside disruptions, and other large scale MHD instabilities, the plasma edge dynamics can be described by a steady state diffusive transport, turbulent transport and the edge-localized modes (ELMs). The latter are short-term break downs of the edge transport barrier, which flatten the pressure gradient and release a certain amount of plasma into the SOL. Depending on the discharge parameters, different types of ELMs are present, generally the larger the frequency the smaller the amount of released plasma. Regimes in which no ELMs occur show better confinement at the expense of impurity accumulation in the core, hence smaller ELMs are considered as a method of releasing both impurities and helium in a burning reactor.

The dynamic loading of the first wall and the divertor by ELMs and micro-turbulence is strongly related to transport and parameters of the transient events. ELMs are known to increase the wall temperature and viewed as a major contributor to wall erosion. For future devices it is imperative to suppress large ELMs altogether since their projections of size and energy content for an operational reactor would erode the surface within days or months. Assuming absence of ELMS, transient effects stemming from micro-turbulence seem unavoidable in fusion reactors. Although transient plasma loads have been simulated by laser irradiation or electron beams [6, 9] the effect of turbulence on the sputtering yield and erosion overall is so far investigated theoretically only [10–12].

The extreme plasma temperature dependence of the sputtering yields requires a high degree of certainty about the temporal temperature evolution, also within the turbulent transport events, for predictions accounting for the gap between time scales of experiments and a reactor are substantial: years of operation in reactors versus hours of sample exposures in current experiments. With a few millimeters of wall thickness to erode from the most common approach to minimize erosion is so far to stay below the sputtering threshold for Deuterium marked in Fig 2.2, which is determined by the erosion yield at the average density and temperature. To understand and evaluate PWI processes governing the turbulent transport and the resulting temperature excursions are reviewed in the following with a focus on similarities between fusion reactors and linear plasma generators.

2.2 Drift Wave Instability and Turbulence

The most common type of waves found in magnetized laboratory plasmas are drift waves, termed by the electron drift velocity with which these waves travel. The associated drift wave instability is also called *universal* instability since it arises from the plasma density gradient, which is present in virtually all bounded plasmas. Since the excited wave modes and the possibly resulting drift wave turbulence plays

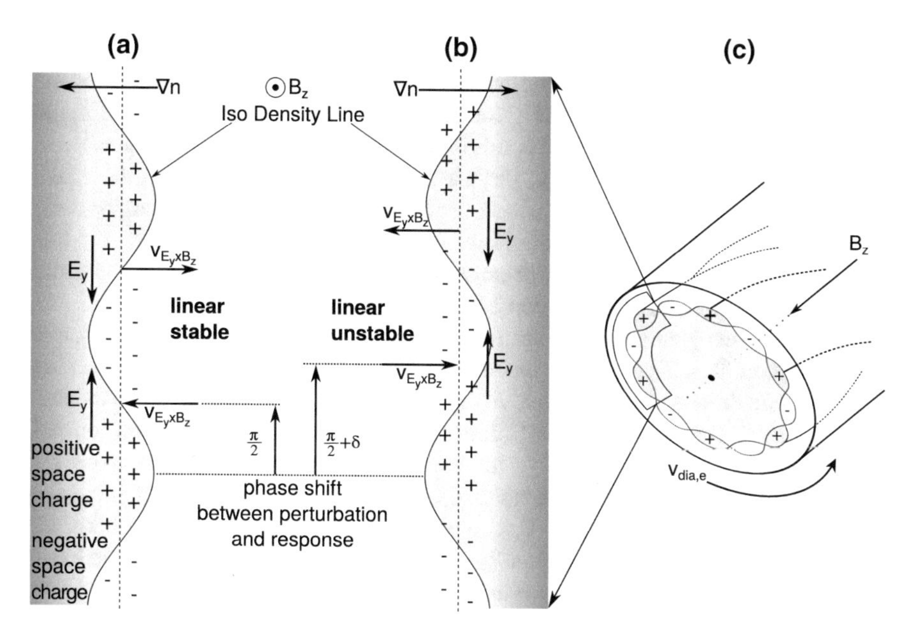

Fig. 2.3 Basic mechanism of stable **a**, unstable **b** drift wave and spatial structure **c** (adopted from [14])

a major role in the edge plasma transport [13] and thus affects the plasma wall interaction, generation and destabilization mechanisms of drift waves are described briefly.

In the most basic description, the Hasegawa-Mima-model [15], the equilibrium density gradient ∇n_0, perpendicular to a homogeneous magnetic field, is disturbed by a small harmonic oscillation $n_1 \sim \exp\{i(k_\perp y + \omega t)\}$. As depicted in Cartesian slab geometry in Fig. 2.3a, the particle excess is countered by the adiabatic response of electrons to the electric potential Φ_1 along the magnetic field according to the linearized Boltzmann relation

$$n_1 = n - n_0 = n_0 \exp \frac{e\Phi_1}{k_B T_e} - n_0 \approx n_0 (1 + \frac{e\Phi_1}{k_B T_e}) - n_0 = n_0 \frac{e\Phi_1}{k_B T_e}. \tag{2.1}$$

Due to their high mobility (low mass) and the assumption of a collisionless plasma, electrons diffuse from the density protrusion and give rise to a negative space charge in the indicated area and vice versa for the remaining excess of ions. This, in turn, creates an electric field E_y which then generates the $E \times B$-flow $v_{E_y \times B_z}$ in the direction of the initial perturbation. Since the response of the electrons is adiabatic, the phase shift between n_1 (and Φ_1) and $v_{E_y \times B_z}$ is 90°

$$v_{E_y \times B_z} = v_{1x} = \frac{E_{1y}}{B_z} = \frac{-ik_y \Phi_1}{B_z} \tag{2.2}$$

and the wave is solely propagating without any growth or dissipation.

In linear plasma devices with cylindrical plasma geometry, the strongest density gradient is azimuthally symmetric (around the z-axis) to first order, which makes $k_y = k_\phi$ and results in boundary conditions for azimuthal modes $m_\phi \lambda_\phi = 2\pi r_m$, where m_ϕ is the mode number, λ_ϕ the wave length of the mode and r_m the radial mode position. In the case of PSI-2, there are even two gradient regions with opposite propagation directions, due to the hollow density profile.

The transition from a single propagating mode to mode destabilization and turbulence requires a delayed phase response of v_{1x} and a non-linear response. To describe the broad frequency range of the plasma edge in tokamaks, the Hasegawa–Wakatani was developed as self-consistent model including resistive effects for drift waves, which include finite parallel conductivity of electrons and thus a mechanism to allow mode destabilization [16]. Processes inducing resistivity include electron-ion collisions, wave-particle interactions or interactions with trapped ion or electrons in a torus or magnetic mirror [13]. As a result, the harmonic oscillation will contain a real part in its argument $\mathrm{Im}(\omega) = \gamma > 0$ with γ as the growth rate of the perturbation.

Generally, these modes grow rapidly in the maximum density gradient and lead via non-linear effects to mode-coupling to excite other mode numbers and a broad frequency range [15]. Especially when the perturbations reach amplitudes in the order of the background field, power transfer to higher harmonics occurs since the approximation in Eq. 2.2 requires additional higher order terms. While this is already possible in the frame of Hasegawa-Mima-model, the growth and redistribution of energy towards interacting modes in the frame of the Hasegawa–Wakatani model gives a more complete picture of plasma transport in the regime of strong turbulence. The free energy of the background density gradient is converted to this broadband spectrum of drift waves until saturation, which occurs eventually upon flattening of the driving gradient.

Besides the above mentioned mechanism to drive drift waves, ion and electron temperature gradients are additional means to drive drift waves [17]. The effect of broadband drift wave turbulence on transport is actually two-fold. Via self-organization, zonal flows are created by drift waves and they can reduce radial transport [18, 19]. On the other hand, a large fraction of the turbulent transport is also generated by drift waves, since they tend to flatten out the gradients essential to the operation of a fusion reactor. Furthermore, even when zonal flows and the resulting transport barriers reduce the radial transport, drift waves are believed to play a role in the generation of fast radial propagating edge transport events, which are the subject of interest for this thesis and thus explained in the following section.

2.3 Intermittent Edge Plasma Transport

Turbulent cross-field transport in magnetically confined plasmas has been an obstacle for the performance of all types of fusion reactors since the very beginning of fusion research and its understanding has grown steadily over the past decades. The complex dynamics in the plasma edge are driven by density and temperature gradients, which are necessary for maximizing the number of fusing particles in the core, while allowing solid wall materials at the edge. Moreover, gradients and shape of the magnetic field and the subsequent particle trajectories lead to additional instabilities, which increase transport and can terminate the plasma discharge in a disruption. As these limitations on achievable confinement require larger reactor volumes with more energy stored in plasma and magnetic field, disruptions and large scale turbulence became a major threat to the operation of the reactor, especially the first wall. Revised reactor geometries, damage intervention mechanisms and additional corrective magnetic fields addressed the problems and provide the necessary measures for a baseline operation. The uncertainties in connection to turbulent transport in the plasma edge and thus the long-term stability of the first wall is now determined by the accuracy of various scaling laws, predicting the reactor performance and behavior towards the ITER scale based on previous machines, but limited physical understanding.

Initially, the transport paradigm in the edge of a tokamak reactor stipulated that the main fraction of the power and particle load crossing the LCFS is transported swiftly down towards the divertor, since fast transport along the magnetic field lines prevents the cross-field transport to the first wall. Unfortunately, measurement campaigns with increasing heating power and density showed a flattening of the pressure profiles in the SOL and thus pointed towards an increasingly convective transport [20, 21]. It turned out that the large pressure fluctuations, formerly named blobs for their characteristic shape in time traces, are actually field-aligned, elongated and coherent structures, which dominate the total amount of transport [22–24]. The convection and filamentary transport were not connected at first, but a detailed analysis of filaments showed their share of transport to be significant or even dominating the total cross-field transport. The previously termed anomalous transport, which was described by a large diffusion coefficient with unclear reason is now necessarily replaced by a paradigm of turbulent transport, composed of snort intermittent filaments. The intermittency, i.e. the absence of a characteristic frequency component in spatiotemporal vicinity of filament occurrence, prevented an immediate understanding of its generation process, while statistical properties were readily available from diagnostics with high resolution. Time traces of plasma pressure in the SOL feature sharp positive spikes, which are caused by large fluctuation amplitudes with a non-Gaussian distribution function. The skewness S and kurtosis K are the third and fourth order moment of a time series x_i with mean x_m and standard deviation σ, and are widely used to describe the fluctuation statistics in the plasma edge:

$$S = \frac{\Sigma(x_i - x_m)^3}{\sigma^3} \qquad K = \frac{\Sigma(x_i - x_m)^4}{\sigma^4} - 3 \tag{2.3}$$

Despite the different magnetic structure in tokamaks, stellarators or linear devices, similar profiles for skewness and kurtosis are found. Similarly, the probability density function, i.e. the amplitude distribution, of plasma fluctuations match for different magnetic confinement devices [25]. Hence the plasma environment creating filaments depends on relative changes in plasma properties and dimensionless scale lengths,

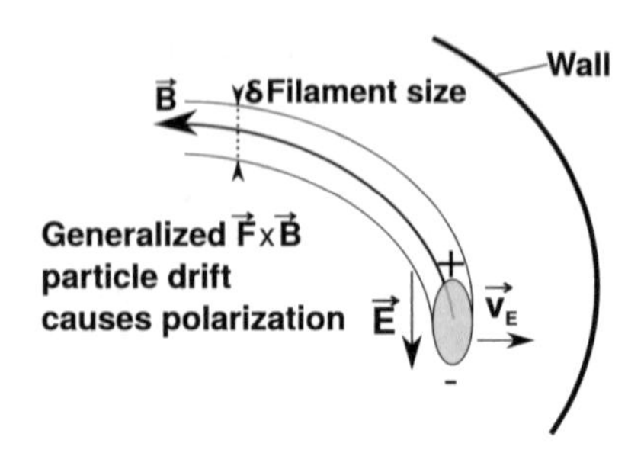

Fig. 2.4 Schematic view of filament structure and propagation mechanism adopted from [26]

which seem to be reached in many magnetically confined plasmas. Nevertheless, the individual mechanisms in a specific environment differ and thus theoretical descriptions for the tokamak SOL filaments might not be applicable for linear plasma generators. While the generation of filaments in tokamaks is linked to magnetic curvature effects [26–28], linear magnetic topologies require drift wave turbulence or shear flows for promoting filament generation in the absence of magnetic curvature [29–31]. Generally, the field-aligned shape depicted in Fig. 2.4 induces a particle drift leading to a polarization. Thus, the propagation of filaments is again caused by different mechanisms leading to self-generated $E \times B$-motion [32], in which the collisionality plays an important role to distinguish between propagation modes [26, 33].

The essence of the extensive body of research in terms of filament structure and motion is the ubiquitous existence of ballistic, coherent transport phenomena, which carry heat and particles across the SOL. Their large pressure amplitudes cause increasing values of skewness and kurtosis in the SOL, depending on the ratio of filament to background pressure, and the transition of the skewness profile to negative values is considered as indication of the filament generation region, which is generally at the position of transition from edge to SOL. The structure size is up to several meters along the magnetic field and in the order of one to a few cm in poloidal (azimuthal) and radial direction, while the radial velocity reaches up to several km/s. Size and velocity are rather influenced by scale lengths and dimensionless parameters than machine size, thus the spread of figures is limited. The absence of a singular theory for filament dynamics leaves only the consistent statistical properties as possibility to judge the influence of filamentary transport in a particular plasma environment, independent of the magnetic configuration. The skewness is especially important as it directly describes the pressure fluctuation distribution and thus the potential influence on PWI processes.

The biggest uncertainty for estimating the influence of filament transport on erosion is the resemblance of the fluctuation statistics to actual temperature fluctuations, since a pressure increase could stem from increased density and/or temperature. While only the latter would lead to enhanced erosion compared to mean value predictions, higher than background temperatures are likely since intermittent structures are generated and ejected from a temperature gradient region. However, secondary effects can also occur if the higher short-term density of plasma impinging on the wall leads to a higher release of impurities, which would otherwise be bound on the surface at an average plasma flux. Thus, besides the direct interaction of filaments with the first wall, the increased recycling of plasma at the first wall is another important implication of convective transport, since it provides a source for impurities and CX processes.

2.4 Diagnostics for Edge Plasma Turbulence

The fluctuations generating turbulence and intermittent filamentary transport are visible in density, temperature, electrostatic potential and velocity fields of the plasma. Ideally, these quantities are measured with spatiotemporal resolution in the range of millimeters and microseconds.

The most commonly used diagnostic for localized measurements are Langmuir probes, inserted dynamically or stationary in the plasma edge of toroidal devices or the whole plasma in basic (linear) plasma devices, where the heat load is usually tolerable. One-point fluctuation measurements give access to the particle flux $\Gamma_e \sim n_e\sqrt{T_e}$, consisting of electron density n_e and temperature T_e, or floating potential V_{fl}, while two-point measurements combine both potentials to infer the local radial or poloidal (azimuthal) electric field, $\vec{E}_r$ and $\vec{E}_p$, respectively. Subsequently, the velocities correspond to $\vec{v}_{r/p} = \vec{E}_{p/r} \times \vec{B}/B^2$), while velocities are also estimated by the delays between (coherent) signals. Larger structural properties can be obtained by 2D probe arrays and combinations of fixed and movable probes. Virtually all plasma devices studying filaments use Langmuir probes, while an extensive characterization of plasma fluctuations was performed, e.g. in TORPEX [34], CSDX [35] or VINETA [36]. Depending on the experimental effort velocity fields and potential structure of filaments can be obtained with probe arrays and by repeated measurements in similar discharges.

However, the incomplete theory about probes in magnetic fields and possible plasma perturbations by the probes themselves make the quantitative interpretation difficult. Active optical diagnostics using a local source of radiation by neutral atom beams directed through the plasma edge are called beam emission spectroscopy and have been used in TEXTOR [37], DIII-D [38] or AUG [12, 39].

Here, only the density profile evolution is measured, while imaging the atomic emissions of an injected neutral gas puff (Gas puff imaging - GPI) produces 2D evolutions of density for a limited time, but both methods achieved resolutions near $\sim$1 mm and $\sim$1 μs [40, 41]. Both methods rely on models to calculate the plasma density from local atomic emissions. Furthermore, the application of fixed frequency reflectometry (FFR) was extended to study filaments at AUG [42]. While the spatial resolution is only in the range of a few cm, the temporal resolution of 0.5 μs enabled radial velocity estimates comparable to Langmuir probe measurements.

Fast visible imaging can provide high spatiotemporal resolutions for filament investigations, if the source of photon emission is strong and localized, e.g. the neutrals in the SOL are only excited by filaments (MAST [43, 44], Alcator C-MOD [45], QUEST [46]). Image analysis provide shape, velocities and fluctuation statistics of filaments, while the plasma parameters causing the emission fluctuations are usually assumed to be density related. Cross-checking of several methods in the same plasma volume confirmed similar fluctuation statistics in a number of experiments [38, 47], however especially disentangling density and temperature fluctuations (e.g. in a collision-radiative model in Helium [48]) is complicated since Langmuir probes and optical methods can not measure both quantities independently (with high temporal resolution).

Thomson scattering measures density and temperature intrinsically independent by laser Doppler broadening and is routinely installed on most toroidal devices, while basic plasma experiments rarely use it, due to the usually lower plasma densities and thus low signal level. If a single laser shot suffices for a measurement, the time resolution is usually $\sim$10 ns (Q-switched Nd:YAG) with a spatial resolution in the range of mm to cm. A 2D density and temperature profile measurement of filaments was realized with five simultaneous laser beams through the edge of AUG, where fluctuation statistics and amplitudes for filaments were investigated [49]. Plasma densities below $\sim 10^{18} - 10^{19}\,\mathrm{m}^{-3}$ usually require multiple laser shots to reach a signal-to-noise ratio (SNR) for a statistically sound data interpretation, though eliminating temporal resolution.

There are a number of additional diagnostics complementing the aforementioned ones, which are usually or intrinsically without time resolution, but modeling or chopping data collection enables drawing conclusions for filaments. Whether the propagating filaments carry hotter than background ions with them can be tested with retarding field analyzers, which intend to measure the ion velocity distribution function (IVDF) by collecting ions in a probe with a voltage-swept grid arrangement [50]. Laser induced fluorescence (LIF) is another laser-based, active spectroscopy, probing the IVDF and requires accumulation of several laser shots. However, the time evolution of the IVDF for an unstable magnetized plasma column was retained by in-phase triggering of laser and plasma rotation [51]. This phase-locking technique is also applicable for TS for a wide range of plasma phenomena [52, 53], since the time resolution is ultimately only restricted by the laser shot duration.

References

1. Loarte A, Lipschultz B, Kukushkin AS et al (2007) Chapter 4: power and particle control. Nucl Fusion 47(6):S203–S263
2. Stangeby PC (2000) The plasma boundary of magnetic fusion devices. Institute of Physics Publishing
3. Sharma AS, Aschwanden MJ, Crosby NB et al (2016) 25 years of self-organized criticality: space and laboratory plasmas. Space Sci Rev 198(1–4):167–216
4. Federici G, Skinner CH, Brooks JN et al (2001) Review: plasma-material interactions in current tokamaks and their implications for next step fusion reactors. Nucl Fusion 41(12):1967–2137

5. Buzi L, De Temmerman G, Unterberg B et al (2015) Influence of tungsten microstructure and ion flux on deuterium plasma-induced surface modifications and deuterium retention. J Nucl Mater 463:320–324
6. Wirtz M, Bardin S, Huber A et al (2015) Impact of combined hydrogen plasma and transient heat loads on the performance of tungsten as plasma facing material. Nucl Fusion 55(12):123017. (article id. 123017)
7. Eckstein W (2007) Sputtering yields: experiments and computer calculations from threshold to MeV energies. Springer, Berlin, pp 33–187
8. Kirschner A, Borodin D, Philipps V et al (2009) Estimations of erosion fluxes, material deposition and tritium retention in the divertor of iter. J Nucl Mater 390–391:152–155
9. Huber A, Arakcheev A, Sergienko G et al (2014) Investigation of the impact of transient heat loads applied by laser irradiation on iter-grade tungsten. Phys Scr 159:014005. (article id. 014005)
10. Marandet Y, Nace N, Valentinuzzi M et al (2016) Assessment of the effects of scrape-off layer fluctuations on first wall sputtering with the tokam-2d turbulence code. Plasma Phys Control Fusion 58(11):114001. (article id. 114001)
11. Mekkaoui A, Kotov V, Reiter D, Boerner P (2014) Effect of turbulent fluctuations on neutral particle penetration and charge exchange sputtering. Contrib Plasma Phys 54(4–6):409–414
12. Birkenmeier G, Manz P, Carralero D et al (2015) Filament transport, warm ions and erosion in asdex upgrade l-modes. Nucl Fusion 55(3):033018. (article id. 033018)
13. Tynan GR, Fujisawa A, McKee G (2009) Topical review: a review of experimental drift turbulence studies. Plasma Phys Control Fusion 51(11):77. (article id. 113001)
14. Chen FF (1984) Introduction to plasma physics and controlled fusion. Volume 1: Plasma physics, vol 1, 2nd edn. Plenum, New York
15. Hasegawa A, Mima K (1978) Pseudo-three-dimensional turbulence in magnetized nonuniform plasma. Phys Fluids 21:87–92
16. Wakatani M, Hasegawa A (1984) A collisional drift wave description of plasma edge turbulence. Phys Fluids 27(3):611–618
17. Horton W (1999) Drift waves and transport. Rev Mod Phys 71:735–778
18. Diamond PH, Itoh S-I, Itoh K, Hahm TS (2005) Topical review: zonal flows in plasma—a review. Plasma Phys Control Fusion 47(5):R35–R161
19. Fujisawa A (2009) Review article: a review of zonal flow experiments. Nucl Fusion 49(1):013001. (article id. 013001, 42 pp)
20. Umansky MV, Krasheninnikov SI, LaBombard B, Terry JL (1998) Comments on particle and energy balance in the edge plasma of alcator c-mod. Phys Plasmas 5(9):3373–3376
21. Antar GY, Krasheninnikov SI, Devynck P et al (2001) Experimental evidence of intermittent convection in the edge of magnetic confinement devices. Phys Rev Lett 87(6):065001
22. Carreras BA (2005) Plasma edge cross-field transport: experiment and theory. J Nucl Mater 337:315–321
23. Müller HW, Bernert M, Carralero D et al (2015) Far scrape-off layer particle and heat fluxes in high density - high power scenarios. J Nucl Mater 463:739–743
24. Carralero D, Manz P, Aho-Mantila L et al (2015) Experimental validation of a filament transport model in turbulent magnetized plasmas. Phys Rev Lett 115(21):215002
25. Antar GY, Counsell G, Yu Y, Labombard B, Devynck P (2003) Universality of intermittent convective transport in the scrape-off layer of magnetically confined devices. Phys Plasmas 10(2):419–428
26. Krasheninnikov SI, D'Ippolito DA, Myra JR (2008) Recent theoretical progress in understanding coherent structures in edge and sol turbulence. J Plasma Phys 74(5):679–717
27. Manz P, Carralero D, Birkenmeier G, et al (2013) Filament velocity scaling laws for warm ions. Phys Plasmas 20(10):102307. (article id. 102307, 8 pp)
28. Riva F, Colin C, Denis J et al (2016) Blob dynamics in the torpex experiment: a multi-code validation. Plasma Phys Control Fusion 58(4):044005. (article id. 044005)
29. Kasuya N, Yagi M, Itoh K, Itoh S.-I (2008) Selective formation of turbulent structures in magnetized cylindrical plasmas. Phys Plasmas 15(5):052302. (article id. 052302, 10 pp)

30. Windisch T, Grulke O, Naulin V, Klinger T (2011) Formation of turbulent structures and the link to fluctuation driven sheared flows. Plasma Phys Control Fusion 53(12):124036. (article id. 124036, 9 pp)
31. Xu M, Tynan GR, Diamond PH et al (2011) Generation of a sheared plasma rotation by emission, propagation, and absorption of drift wave packets. Phys Rev Lett 107(5):055003 (id. 055003)
32. Krasheninnikov SI (2001) On scrape off layer plasma transport. Phys Lett A 283:368–370
33. D'Ippolito DA, Myra JR, Zweben SJ (2011) Convective transport by intermittent blob-filaments: comparison of theory and experiment. Phys Plasmas 18(6):060501. (article id. 060501, 48 pp)
34. Theiler C, Furno I, Loizu J, Fasoli A (2012) Convective cells and blob control in a simple magnetized plasma. Phys Rev Lett 108(6):065005. (id. 065005)
35. Müller SH, Theiler C, Fasoli A et al (2009) Studies of blob formation, propagation and transport mechanisms in basic experimental plasmas (torpex and csdx). Plasma Phys Control Fusion 51(5):055020. (article id. 055020, 15 pp)
36. Windisch T, Grulke O, Klinger T (2006) Radial propagation of structures in drift wave turbulence. Phys Plasmas 13(12):122303. (article id. 122303, 7 pp)
37. Huber A, Samm U, Schweer B, Mertens P (2005) Results from a double li-beam technique for measurement of both radial and poloidal components of electron density fluctuations using two thermal beams. Plasma Phys Control Fusion 47(3):409–440
38. Boedo JA, Rudakov DL, Moyer RA et al (2003) Transport by intermittency in the boundary of the DIII-D tokamak. Phys Plasmas 10(5):1670–1677
39. Willensdorfer M, Birkenmeier G, Fischer R et al (2014) Characterization of the li-bes at asdex upgrade. Plasma Phys Control Fusion 56(2):025008. (article id. 025008)
40. Shesterikov I, Xu Y, Berte M et al (2013) Development of the gas-puff imaging diagnostic in the textor tokamak. Rev Sci Instrum 84(5):053501–053501-11
41. Fuchert G, Birkenmeier G, Carralero D et al (2014) Blob properties in l- and h-mode from gas-puff imaging in asdex upgrade. Plasma Phys Control Fusion 56(12):125001. (article id. 125001)
42. Vicente J, Conway GD, Manso ME et al (2014) H-mode filament studies with reflectometry in asdex upgrade. Plasma Phys Control Fusion 56(12):125019. (article id. 125019)
43. Kirk A, Thornton AJ, Harrison JR et al (2016) L-mode filament characteristics on mast as a function of plasma current measured using visible imaging. Plasma Phys Control Fusion 58(8):085008. (article id. 085008)
44. Thornton AJ, Fishpool G, Kirk A, the MAST Team and the EUROfusion MST1 Team (2015) The effect of l mode filaments on divertor heat flux profiles as measured by infrared thermography on mast. Plasma Phys Control Fusion 57(11):115010. (article id. 115010)
45. Grulke O, Terry JL, LaBombard B, Zweben SJ (2006) Radially propagating fluctuation structures in the scrape-off layer of alcator c-mod. Phys Plasmas 13(1):012306. (article id. 012306, 7 pp)
46. Liu HQ, Hanada K, Nishino N et al (2011) Study of blob-like structures in quest. J Nucl Mater 415(1):S620–S623
47. Antar GY, Yu JH, Tynan G (2007) The origin of convective structures in the scrape-off layer of linear magnetic fusion devices investigated by fast imaging. Phys Plasmas 14(2):022301. (article id. 022301, 10 pp)
48. Ma S, Howard J, Thapar N (2011) Relations between light emission and electron density and temperature fluctuations in a helium plasma. Phys Plasmas 18(8):083301. (article id. 083301, 14 pp)
49. Kurzan B, Horton LD, Murmann H et al (2007) Thomson scattering analysis of large scale fluctuations in the asdex upgrade edge. Plasma Phys Control Fusion 49(6):825–844
50. Kočan M, Gennrich FP, Kendl A, Müller HW, and the ASDEX Upgrade Team (2012) Ion temperature fluctuations in the asdex upgrade scrape-off layer. Plasma Phys Control Fusion 54(8):085009. (article id. 085009, 11 pp)

51. Rebont C, Claire N, Pierre T, Doveil F (2011) Ion velocity distribution function investigated inside an unstable magnetized plasma exhibiting a rotating nonlinear structure. Phys. Rev. Lett 106(22):225006. (id. 225006)
52. Biel W, Abou-El Magd A, Kempkens H, Uhlenbusch J (1995) Study of oscillating magnetized hollow cathode arcs by time-resolved thomson scattering measurements. Plasma Phys Control Fusion 37(6):599–610
53. Crintea DL, Luggenhölscher D, Kadetov VA, Isenberg C, Czarnetzki U (2008) Fast track communication: phase resolved measurement of anisotropic electron velocity distribution functions in a radio-frequency discharge. J Phys D Appl Phys 41(8):082003. (article id. 082003, 6 pp)

Chapter 3
Laser Light Scattering as Plasma Diagnostic

Laser scattering on plasmas as an active spectroscopy method has become an integral part of the diagnostics in fusion research [1]. Laser-based spectroscopy provides non-intrusive measurements and time resolutions based on the duration of a laser pulse. Reliable and quantitative techniques are extremely useful, especially in the transition region between plasma and wall, where material and plasma parameters are dynamically changing over several magnitudes, and sophisticated diagnostics are thus required to make long term predictions.

The scattering of electromagnetic waves off charged particle ensembles can be used to gain information about the state of the charged particles [2]. The processes leading from single electrons wave scattering to the desired signal of Thomson scattering and sources of relevant interfering scattering processes are derived briefly and discussed in this section.

3.1 Laser Scattering

The interaction between electromagnetic radiation and charged or neutral particles can be described as a scattering process, if the interacting photons are promptly re-radiated. Since the laser photon energy in this work is far less then the rest energy of the lightest involved particle, the electron, a momentum transfer described by the Compton effect can be neglected, since $hf = 2.33\,\mathrm{eV} \ll m_e c^2 = 511\,\mathrm{keV}$, where h is Planck's constant, f the frequency of the photon and m_e the electron mass. Furthermore, the electron kinetic energy of up to $50\,\mathrm{eV}$ corresponds to a velocity of less than 1% of c, the speed of light, hence relativistic effects for all scatter centers can be neglected with plasma temperatures in the order of a few eV. These assumptions make the interpretation of the scattered light less cumbersome and straightforward.

M. Hubeny, *The Dynamics of Electrons in Linear Plasma Devices and Its Impact on Plasma Surface Interaction*, Springer Theses,
https://doi.org/10.1007/978-3-030-12536-3_3

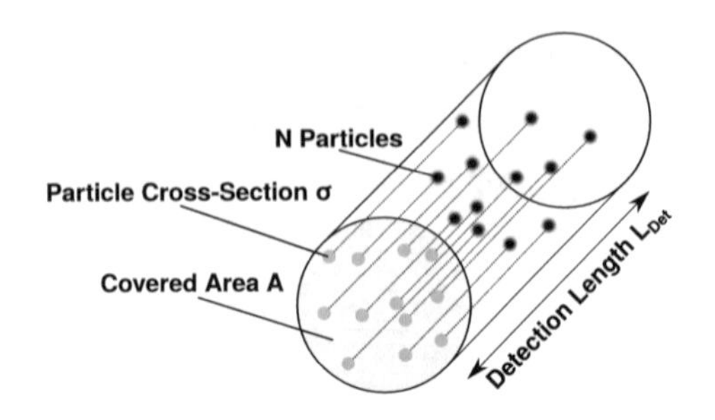

Fig. 3.1 The cross-section and number N of particles in the illuminated area A determine the scattered signal

Therefore, the derivation is presented briefly here and a detailed and comprehensive derivation of laser scattering of charged and neutral particles in plasma is found in Chap. 7 of [3], while an early review for laser scattering as plasma diagnostics tool is reported in [4]. For brevity, key dependencies of several important interactions shall be presented in the following, highlighting their implication towards signal interpretation.

The release of a photon via the acceleration of charged particles by the electric field of a photon is called Thomson scattering and hence the Thomson spectrum contains information about each charged species in a plasma. Neutral particles or molecules interacting with the electric field through the polarizability of their electron clouds is called Rayleigh scattering. Both of these processes are elastic, while in a Raman scattering process rotational or vibrational transitions can be induced in molecules and thus it is an inelastic scattering process.

The fraction of scattered light created by passing through a volume $A \times L_{\mathrm{Det}}$ can be described geometrically by the number of scattering centers as depicted in Fig. 3.1 in terms of density n, (two-dimensional) cross-section σ and length of the scattering volume L_{Det}:

$$\frac{N_s}{N_i} = nL_{\mathrm{Det}}\sigma \ . \tag{3.1}$$

The fraction of scattered and incoming photons, N_s and N_i, can also be replaced by a corresponding power of the signal with $P = Nhf/\lambda$. Since the scattered radiation is generally not isotropic, σ is replaced by

$$\sigma = \frac{d\sigma}{d\Omega}\Delta\Omega, \tag{3.2}$$

where $\frac{d\sigma}{d\Omega}$ is the differential cross-section and $\Delta\Omega$ is the solid angle captured in the experiment.

Macroscopically, the velocity distribution of a plasma species in the plane of the electric field induces a Doppler broadening of the incoming light. Individually, each particle is in motion and thus the scattered photon experiences a Doppler frequency shift in the incoming direction $\omega_p = \omega_i - \vec{k_i} \cdot \vec{v}$ and a subsequent shift $\omega_s = \omega_p + \vec{k_s} \cdot \vec{v}$ in the scattered direction. The wave vectors $\vec{k_i}$ and $\vec{k_s}$ define the scattering plane and point into the direction of incident and scattered photon, respectively. The scattering vector $\vec{k}$ defines the final direction in which the particle velocity is projected upon.

Besides the angular distribution for evaluating the signal strength in a given scattering geometry, the spectral distribution is one of the key parameters sought after in this experiment. Due to their low mass compared to ions, free electrons are accelerated easier and have higher thermal velocities, which increases both signal strength (scattering cross-section) and the Doppler shift. The scattered signal power P_s is derived by the equation of motion for an electric charge in the incident, polarized electric field E_i with negligible magnetic field effects and leads to a spatial power distribution of

$$\frac{dP_s}{d\Omega} = r_e^2 \sin^2 \phi c \epsilon_0 |E_i|^2, \tag{3.3}$$

where ϵ_0 and ϕ are the dielectric constant and ϕ the angle between $\vec{E}_i$ and the scattering plane, respectively, and

$$r_e = \frac{e^2}{4\pi\epsilon_0 m_0 c^2} \tag{3.4}$$

is the classical electron radius with the rest mass being the electron mass $m_0 = m_e$ and e being the electron charge. In accordance to the dipole emission distribution, the power emitted in the direction of acceleration vanishes ($\phi = 0$). The angular dependence of Eq. 3.3 on the angle between $\vec{E}_i$ and the scattering plane is omitted here for simplicity, since it is negligible by using a perpendicular scattering setup. Then, the power distribution normalized by area leads to a differential and total cross-section

$$\frac{d\sigma_{\mathrm{TS}}}{d\Omega} = r_e^2 \qquad \sigma_{\mathrm{TS}} = \frac{8\pi}{3} r_e^2. \tag{3.5}$$

The scattering geometry with the aforementioned electric field and wave vectors is sketched in Fig. 3.2, for the case of a perpendicular setup. The angle θ is defined

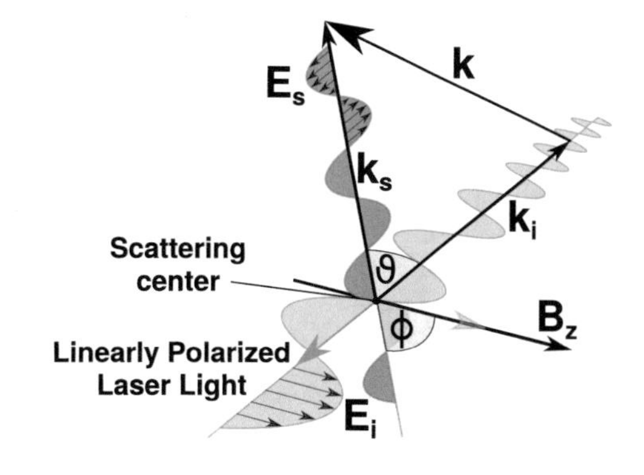

Fig. 3.2 Scattering geometry of wave vectors and electric field for a perpendicularly scattered wavefront from linearly polarized, incoming laser light

by the incoming (red) and the observed (blue) wave direction and will vary slightly depending on the observation trajectory, while ϕ can be chosen strictly perpendicular to maximize the cross-section.

The magnitude of this cross-section is even for electrons rather small with $\sigma_{\mathrm{TS}} = 6.65 \times 10^{-29}\,\mathrm{m}^2$. Therefore, an intense light source is required to generate a sufficient and measurable number of scattered photons. The ion contribution is present as well, but in close spectral vicinity to the Rayleigh signal of neutrals due to the small Doppler shift. Nevertheless, the electrons can carry information of the ion population, depending on the phase relation between all the scattered photons within the excited area and a volume constructed by the Debye length λ_{Debye}, the Debye sphere. The measure of the collectiveness is called scattering parameter [5]

$$\alpha = \frac{1}{k\lambda_{\mathrm{Debye}}} \approx \frac{\lambda_i}{4\pi \sin(\theta/2)} \sqrt{\frac{e^2 n_e}{\epsilon_0 k_{\mathrm{B}} T_e}} \tag{3.6}$$

and relates the electron response to an external field with the scattering wave vector k to a dimensionless number. For the case $\alpha \gg 1$ the electrons in a Debye sphere experience the same phase of the electric field as illustrated in Fig. 3.3b, which is called collective or coherent Thomson scattering. Collective TS is also called ion TS since the collective motion of electrons over a larger scale (than the Debye length) involves the ions as the slow and heavy background charge. In contrast, $\alpha \ll 1$ means the Debye shielding cannot follow the electric field, such that the scattered photons bear no phase relation as the wavelength is much smaller than the Debye length, depicted in Fig. 3.3a. Therefore, this limit is called incoherent TS and contains only information about electrons. For a maximum of density of $n_e = 10^{19}\,\mathrm{m}^{-3}$, a minimal temperature of $T_e = 2\,\mathrm{eV}$, perpendicular scattering $\theta \approx 90^\circ$ and the laser wavelength of $\lambda_i = 532\,\mathrm{nm}$ the scattering parameter is at the largest $\alpha = 0.018$. However, by choosing an observation at a smaller angle the scattering wave vector k decreases towards a more collective photon emission, independent of the plasma parameters. In this work the incoherent TS limit applies to all experimental conditions, since the observation angle strictly perpendicular to the wave vector and densities and temperatures are such that $\alpha \leq 0.018$.

With the approximations above, the individual photons are scattered according to Eq. 3.3 and without collective effects the total power is just the sum of all photons scattered by electrons. The total cross-section and its dependence on the scattering

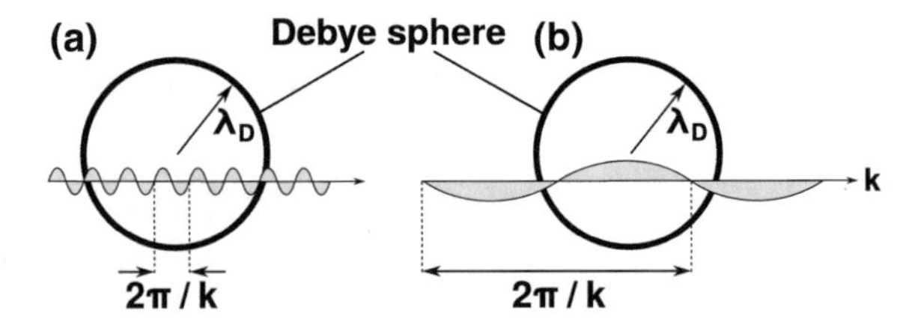

Fig. 3.3 Wave vector and wavelength $\lambda = 2\pi/k$ shown in relation to Debye sphere for **a** incoherent TS and **b** collective/coherent TS

direction is contained in Eq. 3.5, which leads to a relation between scattered photons and the density of the plasma in Eq. 3.1. However, the spectral shape of the scattered photons is caused by the individual velocity of the electrons.

The observed spectral shape can be described mathematically in terms of the electromagnetic radiation in the far field, with linear, non-relativistic motion of the electrons. The scattered electric field as a function of frequency can be written as (cf. [3])

$$\vec{E}_s(f_s) = \frac{r_e \exp(i\vec{k}\cdot\vec{r})}{r} 2\pi\, \vec{e}_s \times \vec{e}_s \times \vec{E}_i\; \delta(\vec{k}\cdot\vec{v} - \omega)\,, \tag{3.7}$$

in which r is the distance to the scattering volume and $\vec{e}_s \times \vec{e}_s \times \vec{E}_i$ resembles the angular dependence. The δ-function in the last term stems from the monochromatic nature of the incident wave and directly projects the velocity field on the direction of $\vec{k}$. Similar to Eq. 3.3, the spatial and spectral power distribution then reads

$$\frac{d^2 P_s}{d\Omega df_s} = r_e^2 \sin^2\phi c\epsilon_0 \frac{P_i}{A_b}\delta(f_s - f_d), \tag{3.8}$$

where $\frac{P_i}{A_b}$ is the mean incident power density and f_s and f_d are scattered and Doppler-shifted frequency, respectively. Integration in velocity space and the number of particles leads to the total energy scattered by n electrons with a velocity distribution $f(\vec{v})$.

The combination of monochromatic input wave and the single frequency scattering field per projected velocity leads to the direct proportionality of the measured spectrum and the velocity distribution in the dimension of $\vec{k}$. Non-thermal features would therefore be directly visible, given a high enough spectral resolution and measurement in the direction in which the depleted or enhanced velocity distribution occurs. However, conclusions about the three dimensional velocity distribution and thus the energy distribution of the electrons are only possible for isotropic distributions.

To evaluate the complete spectrum of the scattered photons for the case of a Maxwellian distribution the spectral dependence of the cross-section is written as

$$\sigma(\vec{k}, \Delta\omega) = \sigma_{\mathrm{TS}} S(\vec{k}, \Delta\omega) = \sigma_{\mathrm{TS}} S_k(\Delta\omega), \tag{3.9}$$

where $S(\Delta\vec{k}, \omega)$ is the dynamic form factor and $S_k(\Delta\omega)$ the projection along $\vec{k}$. For incoherent scattering and a monochromatic light source the projection of $S_k(\Delta\omega)$ is the spectral distribution function. Since the frequency shift is directly proportional to the wavelength shift and particle velocity $\Delta\omega \sim \Delta\lambda \sim v_p$, the probability of finding scattered photons in a wavelength interval $S(\Delta\lambda)d\lambda_s$ follows directly the velocity distribution function. For a one-dimensional Maxwellian distribution of electron velocities the spectral shape becomes

$$S_k(\Delta\omega)d\omega_s = S_k(\Delta\lambda)d\lambda_s = \frac{1}{\Delta\lambda_{1/e}\sqrt{\pi}}\exp\left[-\left(\frac{\Delta\lambda}{\Delta\lambda_{1/e}}\right)^2\right]d\lambda_s, \tag{3.10}$$

where the normalization prefactor

$$\Delta\lambda_{1/e} = \lambda_i\ 2\sin(\theta/2)\sqrt{\frac{2k_B T_e}{m_e c^2}} \tag{3.11}$$

is the width of the scattering spectrum and leads to the electron temperature in the plasma. Solving for T_e leads to

$$T_e = \frac{m_e c^2}{8k_B \sin^2(\theta/2)}\left(\frac{\Delta\lambda_{1/e}}{\lambda_i}\right)^2 . \tag{3.12}$$

Plugging in the used laser wavelength of $\lambda_i = 532$ nm, perpendicular scattering geometry ($\theta = 90\,°$C) gives

$$T_e = 0.452\ (\Delta\lambda_{1/e})^2\quad [\text{eV}], \tag{3.13}$$

with $\Delta\lambda_{1/e}$ given in nm. A schematic depiction of the spectral shape is shown in Fig. 3.4, in which the width and area of the spectrum correspond to the respective electron temperature and density of a Maxwellian plasma. The low intensity ends of the spectrum have to be discriminated against Poisson (shot) noise, while in the central part the superimposed stray-light originates from unwanted laser scattering or other processes explained in the following section.

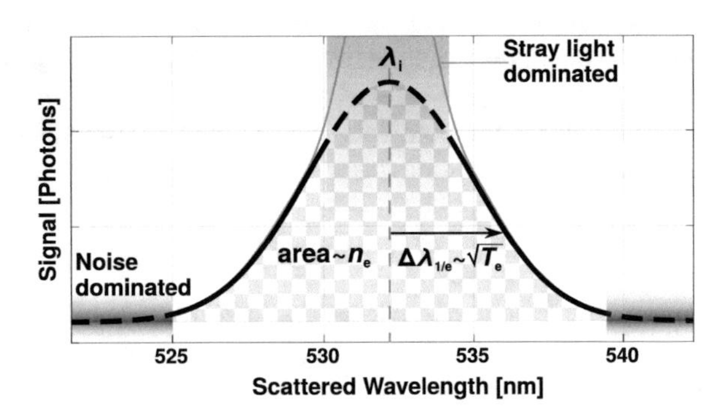

Fig. 3.4 Schematic spectrum for incoherent Thomson scattering with width and area of the Gaussian shape indicating electron temperature and density, respectively

3.1.1 Rayleigh and Raman Scattering

Laser light not only interacts directly with charged particles but also with neutral atoms and molecules by the polarizability of the electron clouds. Elastic scattering with particles much smaller than the wavelength of the incident light is called Rayleigh scattering. For the probability or signal strength of the scattering process Eq. 3.1 applies but with the corresponding scattering cross-sections. These processes can be useful for an absolute calibration of the photons received by Thomson scattering, since the efficiency η of the optical setup is the same for both scattering processes. A detailed knowledge about the Rayleigh and Raman spectrum is necessary to effectively use them as a calibration basis. Serendipitously, these methods are widely used, especially Raman scattering (RS) for its spectrum contains information about chemistry of the probed molecules. In this work, Raman scattering is routinely used, while Rayleigh scattering has been tested as well. Raman scattering has, due to its inelastic character, the advantage of a wider spectrum than Rayleigh scattering and hence occupies a similar wavelength range compared to the Thomson spectrum. The disadvantage of RS having a cross-section 5–6 magnitudes smaller than TS is overcome in practice by filling the chamber with a sufficient neutral pressure. Noble gases are not suitable for RS since it needs molecule transitions and thus at least diatomic gases have to be used with a pressure in the millibar range to provide a particle density in the order of 10^{24}–$10^{25}\,\mathrm{m}^{-3}$. Nitrogen is commonly employed for TS experiments in low temperature plasma experiments [6–8], while Deuterium is preferred for high temperature experiments [9] for its higher transition energy (and thus spectral distance).

The spectrum created by RS contains multiple lines representing the transitions induced during the inelastic scattering process. The two inelastic processes shifting the incident wavelength are rotational and vibrational transitions in the molecule, of which the vibrational transition has a much larger excitation energy and is important only at high temperatures. Thus, at room temperature, all molecules are considered to be in the vibrational ground state. The scattering cross-section determines the initial probability of a Raman scattering scattering event similar to Eq. 3.1:

$$\frac{N_s}{N_i} = nL_{\mathrm{Det}}\frac{d\sigma_{\mathrm{RS}}}{d\Omega}, \tag{3.14}$$

with σ_{RS} being the sum of all line transitions in the RS spectrum:

$$\sigma_{\mathrm{RS}} = \sum_J n_J \sigma_{J\rightarrow J'}, \tag{3.15}$$

where n_J is the density of the initial rotational state and determined by the Boltzmann relation $n_J = Q^{-1} g_J (2J+1)\exp\left(\frac{E_J}{k_{\mathrm{B}}T}\right)$ with Q being the sum of all accessible states and g_J a statistical weight factor, which is $g_J = 6$ ($g_J = 3$) for even (odd) J in Nitrogen. The energy of a state is $E_J = B(J/J+1)$ and $B = 2.48 \cdot 10^{-4}\,\mathrm{eV}$ for

Nitrogen molecules. Q is expressed in [10] by

$$Q = \sum_J g_J(2J+1)\exp\left(\frac{E_J}{k_B T}\right) \approx \frac{9k_B T}{B}. \tag{3.16}$$

The differential cross section of an individual transition $\sigma_{J\rightarrow J'}$ for a perpendicular scattering setup is given in [11] as

$$\frac{d\sigma_{J\rightarrow J'}}{d\Omega} = \frac{64\pi^4}{45\epsilon_0^2} b_{J\rightarrow J'} \frac{\gamma^2}{\lambda^4_{J\rightarrow J'}}, \tag{3.17}$$

where γ is the polarizability anisotropy and $b_{J\rightarrow J'}$ are Placzek-Teller coefficients, given by

$$b_{J\rightarrow J+2} = \frac{3(J+1)(J+2)}{2(2J+1)(2J+3)} \tag{3.18}$$

and

$$b_{J\rightarrow J-2} = \frac{3J(J-1)}{2(2J+1)(2J-1)}. \tag{3.19}$$

The value of $\gamma^2 = (0.39510^{-82} \pm 8\%)\,\mathrm{F}^2\,\mathrm{m}^4$ stems from measurements and references within [11] and is interpolated at $\lambda_i = 532$ nm. With all of the above equations a theoretical Raman spectrum for a given temperature and gas density can be calculated. The spectral width of each transition is smaller than the instrumental function, thus the latter is used for the calculation. The comparison between a theoretical Raman spectrum with all individual lines visible and a resulting spectrum with higher instrumental function can be seen in Fig. 3.5. The statistical weight factor causes the switching heights of the transition lines and the asymmetry between the side lobes of the spectrum is due to the temperature dependence of the rotational state density.

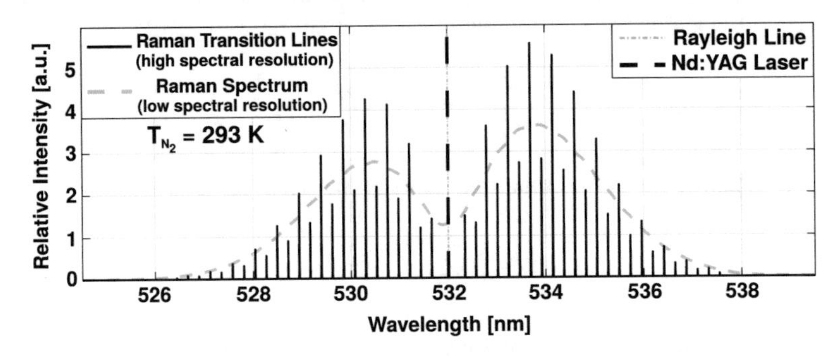

Fig. 3.5 Raman spectrum for Nitrogen at 293 K composed of individual transition lines (black) and the resulting spectrum for finite spectral resolution (dashed gray)

3.2 Laser Absorption

Although a laser with an average power of 10 W might hardly be able to change global plasma characteristics of a 20 kW discharge, a strong laser focus could enable non-linear interactions to become important. A local increase of plasma density and/or temperature through absorptions changes both probing beam and measured quantity and invalidates the measurement. The main absorption mechanisms are inverse bremsstrahlung and photo-ionization [12, 13].

Electron heating by inverse bremsstrahlung is the absorption of a photon by an electron in the vicinity of an ion to conserve the momentum. The energy is deposited in the plasma through collisions during the acceleration in the radiation field and conducted on the time scale of electron-ion collisions and electron-neutral collisions. The maximum temperature increase occurs when the laser pulse is short against these times, which is the case for all plasma conditions in PSI-2. The relative increase of electron temperature has been estimated by [13]:

$$\frac{\Delta T_e}{T_e} = 3.38\ 10^{-39} \frac{n_{\mathrm{ion}} Z^2}{(k_{\mathrm{B}} T_e)^{3/2}} \lambda_i^3 \left[1 - \exp\left(-\frac{h f_i}{k_B T_e} \right) \right] \frac{E_i}{\pi r_{\mathrm{Laser}}^2}, \tag{3.20}$$

where n_{ion} is the ion density, Z is the ion charge number, f_i is the laser frequency, E_i is the laser pulse energy and r_{Laser} is the minimal laser spot radius. The highest density and lowest temperature is assumed together with a diffraction limited laser beam radius at the highest power to estimate the worst case. Taking Z = 1 and thus $n_i = n_e \leq 10^{19}\,\mathrm{m}^{-}3$, $T_e = 1\,\mathrm{eV}$, $r_{\mathrm{Laser}} = 0.3\,\mathrm{mm}$ and $E_i = 1.2\,\mathrm{J}$ yields $\Delta T_e / T_e = 0.3\%$ and decreases rapidly for increasing T_e, hence this effect is negligible.

The absorption by photo-ionization increases the local plasma density and introduces a source of plasma if the ionized plasma states are repopulated during the laser pulse duration. Generally, every electron state requiring less energy than the laser photon provides can be ionized directly, while multi-photon ionization is possible at sufficient power densities. Although the cross-sections are significant for excited Deuterium and Argon states, the state density is low compared to the plasma density since the neutral density in the PSI-2 exposure chamber is strongly reduced by a diaphragm and states close to continuum are easily ionized by collisions already (Boltzmann relation). Furthermore, a strong state depopulation by ionization along the laser path can be excluded as no absorption of laser light was observable within the given 1% accuracy of the laser power measurements.

References

1. Kunze HJ (1968) The laser as a tool for plasma diagnostics. North-Holland Publishing Company, Amsterdam
2. Froula DH, Luhmann NC, Glenzer SH, Sheffield J (2011) Plasma scattering of electromagnetic radiation
3. Hutchinson IH (2002) Principles of plasma diagnostics, 2nd edn. Cambridge University Press (Cambridge Books Online)
4. Evans DE, Katzenstein J (1969) Laser light scattering in laboratory plasmas. Rep Prog Phys 32(1):207–271
5. Salpeter EE (1960) Electron density fluctuations in a plasma. Phys Rev 120:1528–1535
6. de Regt JM, Engeln RAH, de Groote FPJ, van der Mullen JAM, Schram DC (1995) Thomson scattering experiments on a 100 mhz inductively coupled plasma calibrated by Raman scattering. Rev Sci Instrum 66(5):3228–3233
7. van de Sande MJ, van der Mullen JJAM (2002) Thomson scattering on a low-pressure, inductively-coupled gas discharge lamp. J Phys D: Appl Phys 35(12):1381–1391
8. Carbone E, Nijdam S (2015) Thomson scattering on non-equilibrium low density plasmas: principles, practice and challenges. Plasma Phys Control Fusion 57(1):014026. (article id. 014026)
9. Röhr H (1981) Rotational Raman scattering of hydrogen and deuterium for calibrating Thomson scattering devices. Phys Lett A 81(8):451–453
10. Herzberg G (1950) Molecular spectra and molecular structure volume I, 2nd edn. D. Van Nostrand Company Inc., Princeton
11. Penney C, St Peters R, Lapp M (1974) Absolute rotational Raman cross-sections for N2, O2, and CO2. J Opt Soc Am 64(5):712–716
12. Gerry ET, Rose DJ (1996) Plasma diagnostics by Thomson scattering of a laser beam. J Appl Phys 37(7):2715–2724
13. Kunze H-J (2009) Introduction to plasma spectroscopy. Introduction to plasma spectroscopy: Springer series on atomic, optical, and plasma physics, vol 56. Springer, Berlin, n/a 2009. ISBN 978-3-642-02232-6

Chapter 4
Experimental Setup on PSI-2

4.1 PSI-2

The linear plasma generator PSI-2 is operated by the IEK-4 at the Forschungszentrum Jülich GmbH and originates from the upgraded PSI-1, located at the Humboldt University Berlin until 2010. The plasma environment in PSI-2 is utilized to investigate plasma wall interactions and imitate the first wall and especially divertor regions in a fusion reactor [1]. Its steady state plasma operation capability is important for reaching exposure durations and thus fluence levels for relevant scientific insights into fuel retention studies [2] and erosion predictions [3]. Additionally, PSI-2 is thought to be a precursor experiment to a completely remote handled plasma generator JULE-PSI, which is placed in a hot cell to expose activated materials and samples [4].

Figure 4.1 shows a three-dimensional model of the complete experiment on top, while the plasma chamber is sketched with detailed part descriptions in the bottom picture. The plasma at PSI-2 is generated by an arc discharge between a hollow and cylindrical cathode and anode. The Lanthanum hexaboride LaB_6 cathode has an inner diameter and length of 6 cm and 7 cm, respectively, while the anode is made of Molybdenum with an inner diameter and length of 5 cm and 40 cm, respectively. The cathode is heated to about 1600 °C to lower the potential for electron emission towards the anode. The anode occupies the whole length of first and second magnetic coil and is grounded to the chamber walls.

The magnetic field at PSI-2 is generated by six water cooled copper coils, which create a steady state magnetic field of up to $B_z = 0.1$ T in-between the coils and source region and up to 0.3 T in the center of the two main coils enclosing the exposure chamber. The magnetic field lines and strength of the two vector components are visualized in Fig. 4.2a and b, respectively, and are based on an IDL routine using the Biot-Savart law with the standard settings at PSI-2. Following the cathode region in positive axial direction, the magnetic field contracts and intersects the cathodes tubular shape. From its inside the emitted electrons are accelerated towards the anode and ionize the working gas, which is inserted through an inlet between cathode and anode. Since the further contracting magnetic field does not connect the electrodes

M. Hubeny, *The Dynamics of Electrons in Linear Plasma Devices and Its Impact on Plasma Surface Interaction*, Springer Theses,
https://doi.org/10.1007/978-3-030-12536-3_4

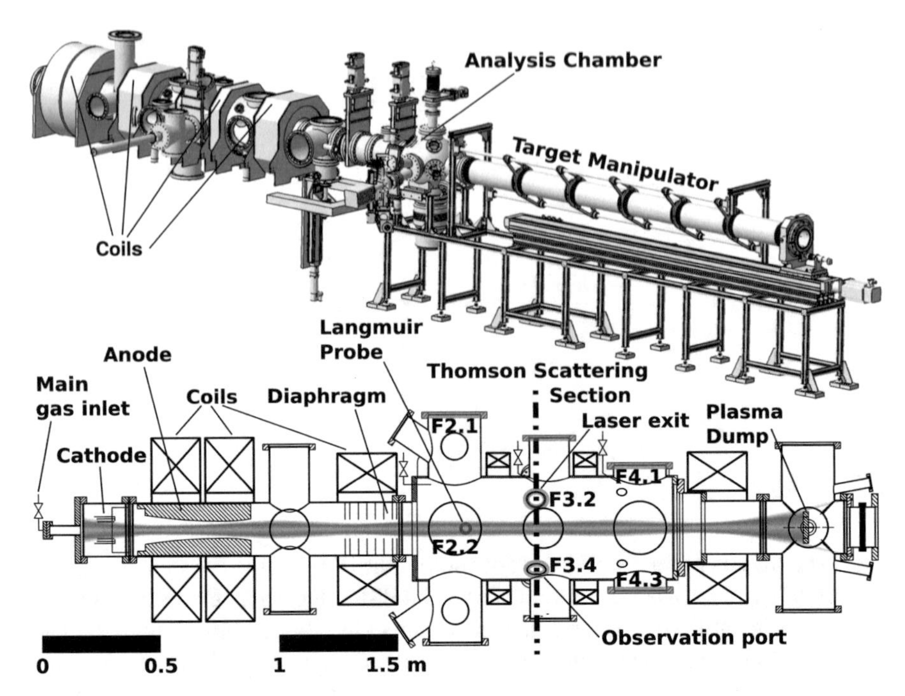

Fig. 4.1 Technical drawing [3] and schematic (with courtesy of C. Brandt, FZJ) of PSI-2 with plasma shown in scale and shape

directly, both plasma and neutral gas are forced to flow through the anode into the magnetized plasma chamber, which consists of 3 parts: Differential pumping stage, exposure chamber and a plasma dump to terminate the plasma column. The first stage is separated from the rest of the chamber by a gas diaphragm, which blocks the neutral gas around the plasma column, thus allowing the second stage to have far less neutrals in the plasma and in the region between plasma and chamber. This increases the ionization degree and gives flexibility in the amount of neutrals desired in the exposure [5].

The shape and diameter of the plasma profile at the source is defined by the strength of the magnetic field downstream as visualized in Figs. 4.1 and 4.2, while the profile broadens by cross-field transport along the axis. The two additional smaller coils in the exposure region generate an almost constant magnetic field, similar for all indicated measurement positions. In contrast to linear devices with a completely homogeneous magnetic field, the changing magnetic field could also modify plasma transport by curvature effects. Since the installed TS diagnostic is mainly compared to the Langmuir probe, their axial distance of roughly 30 cm should result in comparable plasma parameters, aside from the potential influence of the flux expansion behind the diaphragm on the probe measurements.

The first wall of a fusion reactor will have a number of species present in its vicinity. Besides the fuel (DT) and "ash" (Helium), Nitrogen, Argon or other noble

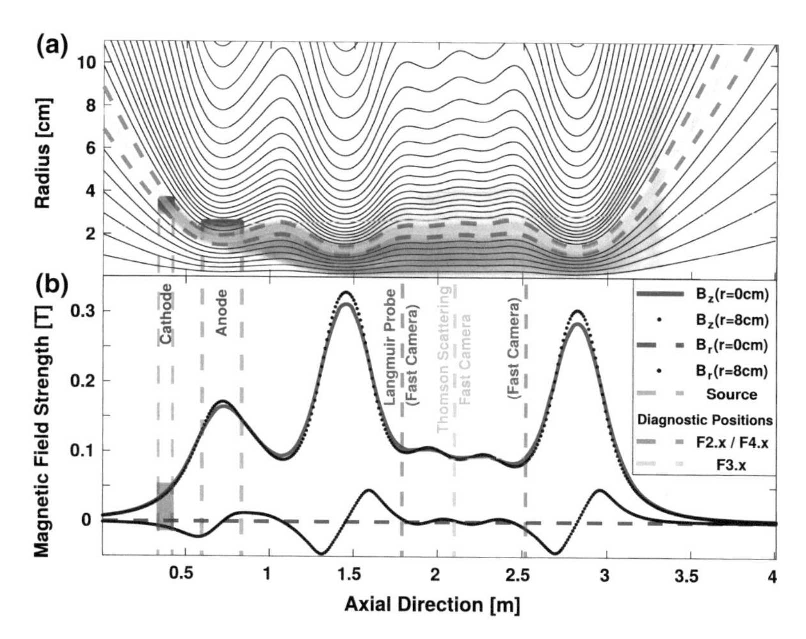

Fig. 4.2 Standard magnetic field configuration at PSI-2. **a** Field lines in azimuthal plane with indicated source position and resulting plasma profile projection along the field. **b** Magnetic field strength versus axial position for radial and axial vector component at center and edge (r = 8 cm) with indicated diagnostic positions

gases might be used to cool down the SOL by radiation. Thus, working gases at PSI-2 include H_2, D_2, He, N_2, Ne, Ar, Xe, Kr, which are commonly introduced into the source, but can also be injected into the differential pumping stage or exposure chamber.

4.2 Langmuir Probe

Langmuir probes are a standard diagnostic tool for low and high temperature plasma, since they are cost effective and versatile. They were first described in the 1920s by Langmuir and Mott-Smith as cylindrical electrode(s) immersed into plasma. Single or multiple electrodes, usually made of Tungsten, are biased with a AC or DC voltage source or left unbiased at the so-called floating potential. Depending on the operational mode the basic plasma parameters accessible are the density, temperature and floating potential of the plasma at high temporal resolution. Multi-pin and/or multiple probes allow the derivation of parameters like electric fields or spatial correlations within the plasma. Both construction and data analysis of Langmuir probes are covered in great detail in [6], since the exact probe plasma interactions depend on

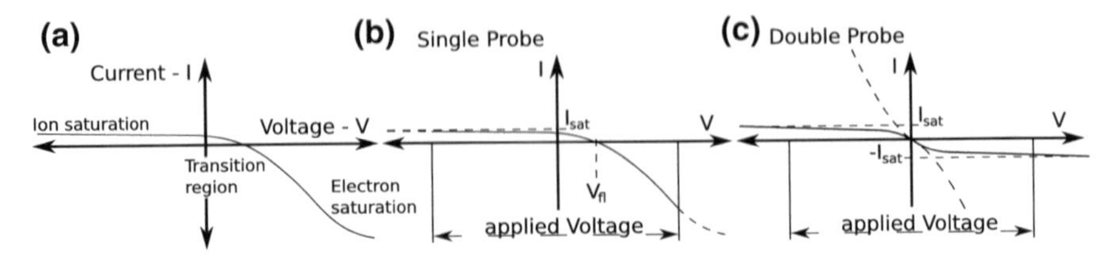

Fig. 4.3 Current response (I-V-curve) for single pin Langmuir probe and combined for a double probe

many aspects of probe and plasma conditions. Only the basic ideas of probe analysis are presented in the following.

For the basic plasma characterization at PSI-2 a double probe design is used, which consists of two identical tungsten pins with 1.5 mm diameter and height, spaced 1.5 mm apart and housed in a ceramic tube. The double probe pins are either biased relative to each other with a floating power supply at an amplitude of 45 V and a frequency of 50 Hz. At the maximum voltage amplitude of each cycle, one pin is collecting ions and the other electrons. Alternatively, only one pin can be used with same power supply but grounded, thus the probe is used as single probe and the second pin can be used floating or biased at a fixed voltage.

To illustrate the current response of a single (b) and double probe (c), a typical cycle of a single probe is shown in Fig. 4.3a, b, where the density defining region is at the ion saturation and the temperature defining region is at the transition. In the ion saturation the bias is high enough to repel all electrons and the current drawn by the probe is given by:

$$I_{\text{sat}} = Cen_e A_{\text{eff}} \sqrt{\left(\frac{k_B T_e}{M_i}\right)}, \tag{4.1}$$

where C is a constant accounting for the presheath density reduction, A_{eff} is the effective current collecting area of the probe and M_i is the ion mass. The ion saturation current in PSI-2 is much smaller than the electron current, since electrons are lighter and thus much faster than ions at comparable species temperatures.

As the negative bias of the probe is reduced electrons start to reach the probe and at equal ion and electron current the floating potential V_{fl} is defined in the I-V-curve. The transition region for a single probe configuration is defined by an exponential increase of the electron current

$$I_e = I_{e,\text{sat}} \exp\left(\frac{-eV_{\text{bias}}}{k_B T_e}\right), \tag{4.2}$$

with increasing bias voltage V_{bias}. If the electron energy distribution is Maxwellian the exponential part is characterized by a temperature. Otherwise, additional information can be extracted from the I-V-curve by its derivatives or simply a logarithmic plot, which would already reveal features like two linear slopes for e.g. a hot and a cold electron population.

Although the electron saturation current is defined similar to the ion saturation, with the corresponding electron mass and effective area, a clear saturation is usually not reached. The mobility of the electrons in the plasma leads to an expansion of the effective probe collection area (sheath expansion), while the heat load by the high electron current causes eventually electron emission. Thus, the exponential increase is usually limited to a non-destructive limit.

The double probe design circumvents this problem by limiting the total current to the ion saturation current, but requires either strictly identical probe tips or the addition of another degree of freedom in the I-V-curve interpretation. Furthermore, plasma gradients between the two probe tips can change the effective voltage. For two probe surfaces A_1 and A_2 the measured current reads

$$I = I_1^{\text{sat}} \left[\frac{\exp\left(\frac{e\, V_{\text{bias}}}{k_B\, T_e}\right) - 1}{\exp\left(\frac{e\, V_{\text{bias}}}{k_B\, T_e}\right) + \frac{A_1}{A_2}} \right], \tag{4.3}$$

where I_1^{sat} is the ion saturation current for tip 1 with the area A_1. The advantage of a potential free measurement with respect to the vessel or plasma is also excluding the measurement of the floating potential. Therefore, both probe configurations are used at PSI-2, depending on which information is required.

The presented equations describing and analyzing the I-V-curves are the most basic and most commonly used ones. Further corrections are made based on the electronic circuit properties of the probe tips and connecting wires. Before plasma operation, the sweeping voltage is applied without plasma and magnetic field to infer resistance and capacitance of an equivalent circuit, which is then considered in the IDL data evaluation program, which is employed for the I-V-curve analysis at PSI-2. Additional corrections based on physical processes at and around the probe surface depend on the plasma regime (collisionality) or type of discharge (DC or RF) and most importantly the presence of a magnetic field [6, 7]. Especially the latter can influence the effective probe surface by limiting it to the projected area parallel to the magnetic field, while it further depends on the collisionality how strong this area restriction takes effect. Although the Debye length and the electron Larmor radius are generally smaller than the probe dimensions, especially the heavier ion species (Argon, Krypton) can be considered unmagnetized due to their larger Larmor radii. Therefore, a precise description of the probe vicinity is rather complex and would need adjustments for every gas type, measurement location (radius) and discharge setting. However, using the basic description in the equations above serves as a good approximation nonetheless and provides a quick estimate of the plasma parameters, while errors from the aforementioned problems must be kept in mind when analyzing probe measurement results.

Besides the absolute measurements of plasma parameters, fluctuations around a mean value contain valuable information about plasma dynamics. To gather information about the plasma fluctuations the plasma response can be measured with up to 100 kHz on the standard acquisition system, resulting in a Nyquist frequency of 50 kHz and a resolution of 10 μs for transient events. In this configuration only one pin of the probe is biased to −200 V, collecting the ion saturation current. The fluctuations of I_{sat} are caused by density ($\sim n_e$) and temperature ($\sim\sqrt{T_e}$), but the latter are commonly assumed to be negligible. The floating potential is measured with the second pin of the double probe via a high resistance voltage divider. A transient recorder (HIOKI 8861) was available only for selected discharges, in which the spectral range of the plasma response could be extended to 1 MHz.

4.3 Fast Camera

Fast framing cameras are simple to use yet their videos contain useful information about dynamics in the range of several 100 kHz. CMOS pixels with high quantum efficiency and sensitivity collect and convert light with high read out speeds, since the circuits for amplification and digitalizing are already integrated into the pixel area. Naturally, the receivable light per frame decreases as the temporal resolution is increased and a certain upper limit is reached depending on the total brightness of the plasma.

The Phantom V711 was the main camera used to characterize the plasma dynamics, while the somewhat slower Phantom V641 was used supplementary. Main properties and operational limits of both cameras are listed in Table 4.1. The fastest frame rates of 1.4×10^6 and 2.19×10^5 fps can only be achieved at the smallest "Continuously Adjustable Resolution" (CAR). Since the pixels are not binned the imaged area at high speeds decreases. Thus, a high aperture wide-angle lens (f/1.8 18−28 mm) was required to image of the whole plasma diameter plus the edge. The spectral response of both cameras compared to typical line radiation is shown in

Table 4.1 Properties of both fast framing CMOS cameras

Phantom camera	V711	V641
Spectral sensitivity	Monochrome (300–900 nm)	Color
Full resolution	1280×800 pixel	2560×1600 pixel
Pixel size	20 μm	10 μm
ISO sensitivity	20,000T; 6,400D	1,600 D
Bit-depth	8/12 bit	8/12 bit
CAR	128×8 pixel	256×8 pixel
Fasted readout	1.4×10^6 fps	219×10^3 fps
Minimum exposure	300 ns	1 μs
Internal memory	8 Gbyte	32 Gbyte
Timing accuracy	≤20 ns	≤20 ns

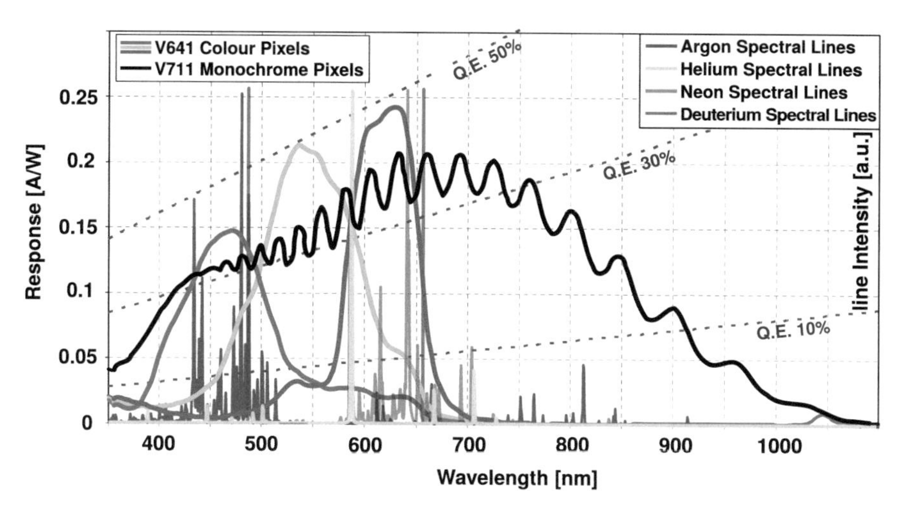

Fig. 4.4 Spectral response of fast cameras (compiled from manufacturer data sheets) with indicated quantum efficiency compared to typical line radiation of Argon and Helium, Neon and Deuterium (only H_α and H_β)

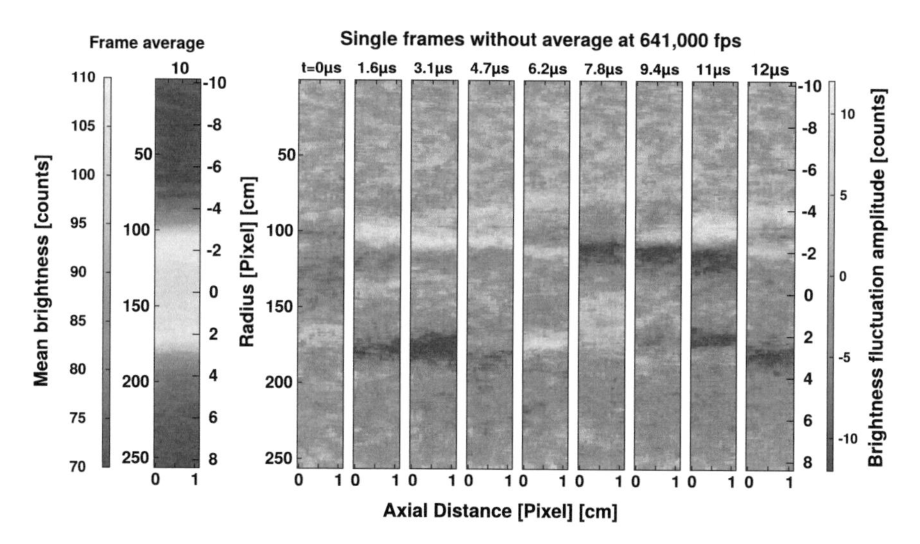

Fig. 4.5 A side view of the frame averaged brightness of a Deuterium discharge (left) and individual camera frames showing the brightness fluctuations around the mean (right). Both spatial scales are marked on either side with corresponding intensity scales

Fig. 4.4. Although filters for all common spectral lines exist, the transmission at the spectral line is on the order of 50%. After verifying a similar frequency response with and without filter (especially for Deuterium), no filters were used to increase the frame rates. A further frame rate gain is envisioned with the nominally 65%

increased transmission of a f/1.4 24 mm lens, while the real improvements should range around 30%, based on the overall transmission of the lens system.

As there are many accessible windows at PSI-2, the cameras were used at side ports (indicated red in Fig. 4.1), at windows F2.1 to F4.3 and the hollow cathode even allowed axial imaging. The raw images captured by the V711 camera are shown in Fig. 4.5, where the average brightness profile is shown with pixel and cm scale on the left. Individual, successive frames with only the fluctuation brightness parts are shown in the right, where radially localized structures are clearly visible.

References

1. Linsmeier C, Unterberg B, Coenen JW et al (2017) Material testing facilities and programs for plasma-facing component testing. Nucl Fusion 57:092012
2. Reinhart M, Kreter A, Buzi L et al (2015) Influence of plasma impurities on the deuterium retention in tungsten exposed in the linear plasma generator psi-2. J Nucl Mater 463:1021–1024
3. Kreter A, Brandt C, Huber A et al (2015) Linear plasma device psi-2 for plasma-material interaction studies. Fusion Sci Technol 68(1):8–14
4. Unterberg B, Jaspers R, Koch R et al (2011) New linear plasma devices in the trilateral euregio cluster for an integrated approach to plasma surface interactions in fusion reactors. Fusion Eng Des 86(9):1797–1800. In: Proceedings of the 26th symposium of fusion technology (SOFT-26)
5. van Eck HJN, Koppers WR, van Rooij GJ et al (2009) Modeling and experiments on differential pumping in linear plasma generators operating at high gas flows. J Appl Phys 105(6):063307–063311
6. Chen FF (1965) Electric probes. Plasma diagnostic techniques. Academic, New York
7. Demidov VI, Ratynskaia SV, Rypdal K (2002) Electric probes for plasmas: the link between theory and instrument. Rev Sci Instrum 73(10):3409–3439

Chapter 5
Thomson Scattering Setup

The development of Thomson scattering as a (plasma) diagnostic was enabled by powerful lasers, first the Ruby laser and then the Nd:YAG laser, predominantly used in high-temperature, high-density plasmas. High efficiency setups with strong stray-light suppression were developed [1] to further extent the application of Thomson scattering towards low-density plasmas or even single shot measurements at higher densities. As a non-intrusive and active spectroscopy method it is considered one of the most accurate diagnostics for plasma parameters [2, 3].

Equipping PSI-2 with a Thomson scattering diagnostic was part of presented work. Therefore, the complete construction and characterization of the light path for the Thomson scattering setup, with the central components being the light source and spectrometer, are described in this section. A particular timing scheme is detailed to provide a basis for the methods allowing time resolved measurements in the following Chap. 6.

5.1 Light Source and Beam Path

The source for the scattered light is the Q-switched Nd:YAG laser Innolas Spitlight 2000 [4] with three amplification stages and a KTP crystal doubling the frequency of the emitted light from 1064 to 532 nm. The Nd:YAG crystals are optically pumped by Xenon flash-lamps with variable intensity. Between the second and third amplification stage the beam is expanded to a 1 cm diameter to achieve a maximum energy of 2 J per pulse at the fundamental frequency, while the maximum energy in the first harmonic is 1.2 J per pulse. The two frequency components are separated by two dichroic mirrors guiding both beams parallel out of the laser, but dumping the fundamental frequency upon the laser exit. The length of a laser pulse at the fundamental frequency is 6–8 ns according to the specifications.

M. Hubeny, *The Dynamics of Electrons in Linear Plasma Devices and Its Impact on Plasma Surface Interaction*, Springer Theses,
https://doi.org/10.1007/978-3-030-12536-3_5

PSI-2 is located in a larger hall on an aluminum platform to maintain access to the cooling circuits below the machine, while the plasma chamber is accessible on top of the platform. The space around PSI-2 for equipment is limited and unsuited for temperature sensitive diagnostics, since the hall is not temperature regulated. Therefore, the laser light is directed to the plasma chamber from a adjacent room via six dielectric mirrors, of which the last three are motorized and remotely controlled. Since the laser path stretches over 30 m, a telescopic expander was used to triple the beam diameter of the laser, which also reduces the divergence by the same factor of three. The seventh and last mirror has a concave surface with a focal length of 1900 mm and consists of a machined copper block with an aluminum coating, reducing the number of optical elements by combining mirror and lens. However, the custom made motorization could not be adopted to the last focusing mirror and would likely be too coarse for beam alignment. Thus, following the beam path design of the TS setup at DIFFER, a more precise piezoelectric control for the last mirror stage would enhance the alignment and stray-light suppression capabilities [5].

Surveillance cameras are used to check the position of the laser beam by a small fraction of its diffuse reflection. To avoid flickering 1% gray filters are used with the highest exposure time setting of the cameras. Figure 5.1 shows the laser spot on the last four mirrors, where only the left is a colored image. The right image contains two laser mirror spots by using a small extra mirror reflecting half of the camera view field. Although the beam alignment was performed manually, the cameras and motorized mirrors already provide the basic hardware for an automated alignment system. This was realized with similar cameras and a software algorithm for the TS system at EAST [6].

Entrance and exit of the laser light to (and from) the vacuum chamber are realized with Brewster windows and a system of baffles in extended tubes guiding the focused laser beam to and from the center of the chamber. Both tubes contain each three baffles with reducing diameter towards the chamber. The window surface inclination of 56.65° is defined by the refractive index of the BK-7 glass and oriented such that a lossless transmission of light occurs for a horizontal polarization (parallel to the magnetic field vector) and thus for the required maximum RS and TS emission perpendicular to initial polarization $\vec{E}_i$ and wave vector $\vec{k}_i$.

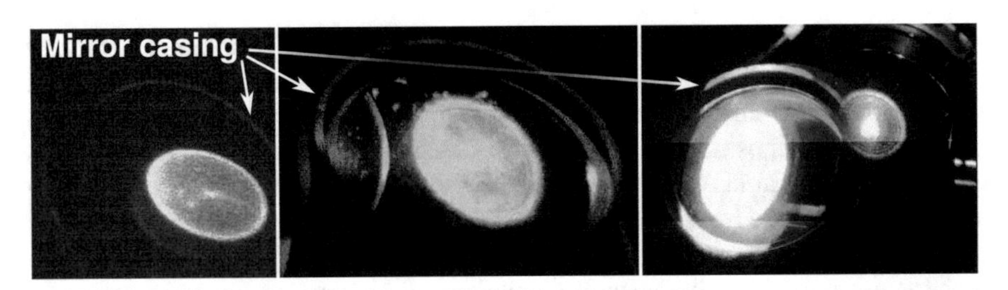

Fig. 5.1 Video feeds of surveillance cameras observing diffuse reflections of the laser spot through gray filters

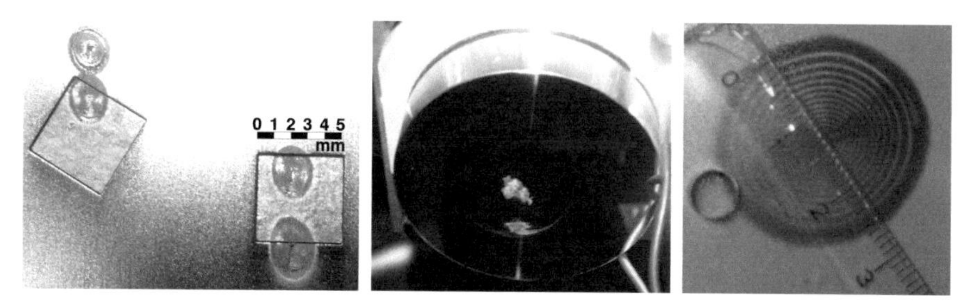

Fig. 5.2 Laser test shots on samples (left), damaged focusing mirror (center) and laser beam profile about 1 m from focusing mirror (right)

The beam dump behind the exit Brewster window is made of numerous small, regular needles compressed in a 10 cm diameter housing in front of a 1% neutral density filter and a white plastic disc (PMMA) of 1 cm thickness. This construction allows only a fraction of the laser light to pass through and visualize the laser spot clearly and safely. The first version of the needle stack suffers from inhomogeneous needle spacing and the diffusing PMMA, restricting the use as a calibration aid. However, with an improved version and employment of another surveillance camera with image processing, the intensity and spot shape of the laser can be utilized for relative laser calibration and beam alignment.

The minimal laser beam diameter at the focus point can be estimated by the divergence and the focal length of the last (concave) mirror. With an initial divergence of 0.5 mrad, the divergence after expansion is 166 μrad at a beam diameter of 3 cm. Constructing the focus of the two wavefronts, which are tilted by ± 166 μrad results in the focal points being -39 and $+41$ mm apart from the undisturbed focus. This results in a minimal spot diameter of $d_{\text{Beam}} = 640\,\mu$mm as an estimate. The real diameter is about three times larger with roughly 2 mm as seen on Fig. 5.2 (left), where samples were shot only about 5 cm out of focus.

Whether the focusing mirror, the beam path or the laser itself caused the increased spot size remains unclear. The beam quality one meter before focusing is seen in Fig. 5.2 (right), showing expected and distinct diffraction rings, which are caused by the beam expander, laser rod aperture and the 30 m beam path [7]. In a future setup, the focusing mirror will be replaced by a conventional mirror and lens or a silver coated focusing mirror, since the aluminum coating failed far below the assumed damage threshold (Fig. 5.2 center). Dust and quality variations could be excluded, since four mirrors showed damages even after extensive dust precautions and reducing the pulse energy to 0.4 J/Shot, corresponding to an average energy density on the mirror of $0.05\,\text{J}\,\text{cm}^{-2}$. Operating the laser at the nominal maximum output for which the internal laser optics are optimized in combination with a more durable lens yields at least two times the usable pulse energy and possibly a better focus.

The beam diameter defines the volume from which all the desired signal is directly emitted and then captured, although scattered light reflected from the wall opposite to the observation can be captured as well. Therefore, a stripe of 28 cm by 5 cm of highly absorbing metal oxide foil (98%,"Spectral Black" by Acktar) was used as a so called view dump, minimizing the reflected light in view of the imaging lens, both from laser scattered and plasma light.

A further reduction of unwanted light from the plasma and laser stray-light can be achieved by a polarization filter on the light collecting lens. Since the TS signal is linear polarized, the relative gain to the plasma background is a factor of two, similar to stray reflections of the laser beam. Unfortunately, standard polarization filters are optimized for maximizing the blocking ratio, but not the transmission, which is only in the order of 50%. Therefore, the provided filter was not usable for settings in which no signal could be sacrificed, especially if plasma background is already low. However, specialized filters are available with nearly 100% transmission for the correct polarization and thus no compromise is needed.

The light path of the complete Thomson scattering setup is shown schematically in Fig. 5.3, where the laser path described above is only outlined in the lower part. The plasma scattering section in PSI-2 is shown on the left and the technical description of the spectrometer (right) follows the scattered light from its origin to the capturing sensor.

5.2 Triple Grating Spectrometer

To analyze the scattered light and discriminate it against stray-light an advanced spectrometer with spatial line filter and gated intensified CCD was tailored to the wavelength range around the central laser line. The spectrometer was designed, built and tested in collaboration with the Ruhr-University-Bochum. The intricate triple grating design ensures an excellent suppression of stray-light by combining three spectrometers in series, one of which is setup as a revertive dispersion element (reversing the dispersion). Naturally, the optical losses triple although the image aberrations of forward and reverse dispersion elements partially (first order) cancel out by using identical holographic gratings and achromatic lenses.

The light enters the spectrometer by a 30 m, 132 optical fiber bundle from CeramOptics. The fibers on the receiving side are arranged linearly and have a core and cladding diameter of 250 and 275 μm, respectively. A 135 mm f/2 tele lens projects an approximately 4:1 demagnified image of a scattering volume with a width of roughly 1 mm by a height of $0.275\,\text{mm} \times 4 \times 132 = 145.2\,\text{mm}$. This ensures covering the whole plasma diameter and a substantial distance towards the vessel walls. At the design distance of 585 mm the solid angle of the image side of the lens matches the solid angle corresponding to the numerical aperture $N = 0.22$ of the fibers.

Theoretically, the spatial resolution of a fiber is $4 \times 0.275\,\text{mm} = 1.1\,\text{mm}$, but since the exit fibers are arranged in a 3 by 44 honeycomb structure (entrance slit in Fig. 5.3), the image is further compressed upon entering the spectrometer and subsequently the spatial resolution is diminished by factor of three. While lowering the spatial and spectral resolution the reduced entrance slit height is beneficial for

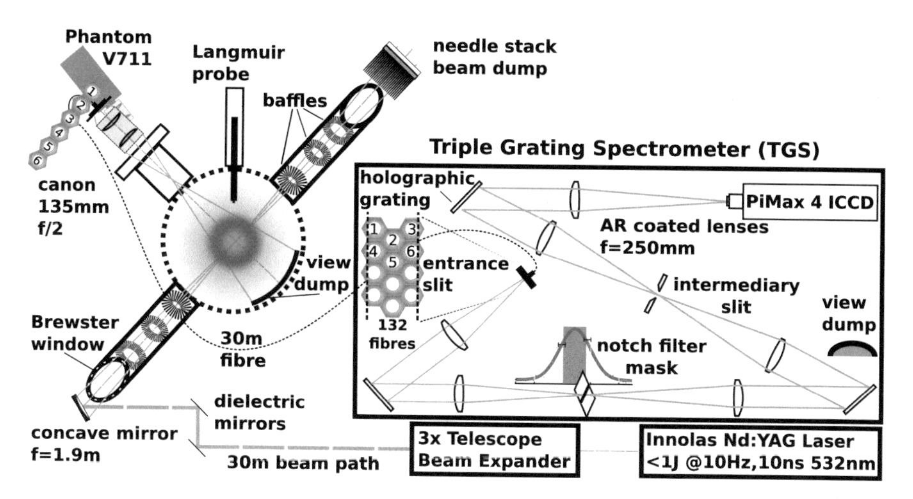

Fig. 5.3 Schematic of laser path, scattering geometry and triple grating spectrometer

minimizing image aberrations and vignetting in the further optical path. A slightly different arrangement of the fibers at the exit towards the spectrometer could improve the spatial resolution by roughly 10–20%, since there is an overlap of fibers which are two places apart (2 and 4, 5 and 7 …), as can be seen in the numbered fiber sorting in Fig. 5.3.

The top right part of Fig. 5.3 shows the light path through the spectrometer and depicts all optical elements. All lenses have a focal length of 250 mm, a diameter of 100 mm and an anti-reflection coating. The distance between lenses and image planes are accordingly dimensioned to capture the full solid angle of all fibers and guide them parallel onto the first holographic grating. The zeroth orders of reflection are absorbed by light dumps (only one shown), while the first order is reflected at 35° towards the first, intermediate imaging lens. Three possible sizes of a notch (1/2/3 mm) filter physically block an area around the central wavelength in this imaging plane - resulting in a superb filter ratio - while the rest of the spectrum passes through a lens onto the second grating operating in reversed dispersion setting. Stray-light generated at and beyond the notch filter generate a small signal at the central wavelength and results in a finite filter ratio of roughly $\sim$1:10^6.

Ideally, refocusing the transmitted light of the second grating results in a 1:1 image of the "entrance slit"-fiber-array while (higher order) image aberrations distort the image. An intermediate slit in this second image plane resembles now the resolution defining element for the following and last grating in the spectrometer. If this intermediate slit is set to less then the width of the honeycomb fiber structure, substantial amounts of photons are lost but a precise wavelength calibration is possible. Furthermore, allowing only the central row of fibers recovers the spatial resolution of a single fiber. The ideal width to capture all scattered light without stray-light is

0.75 mm but image aberrations and imperfect alignment increase this width to about 1 mm used during most experiments. In this setting the remaining light is transmitted to the last grating and imaged as spectrum without the blocked wavelength region onto the camera chip.

5.2.1 *iCCD Camera*

The camera used to capture the spectra produced by the spectrometer is a Princeton Instruments Pi-Max 4 camera [8] with an intensified CCD sensor, which consists of a front-illuminated 1024 × 1024 pixels CCD sensor and a fiber-coupled Gen III intensifier to increase the signal generated by the incoming light. Incoming photons are first converted to electrons (pe^-) by the photo-electric effect in a (negatively biased) photo-cathode and are then accelerated towards a micro channel plate (MCP), which is a matrix of electron multipliers with an amplification factor depending on the applied voltage. The gating of the camera is realized by combining the bias of the photo-cathode and the MCP voltage. The multiplied electrons are finally accelerated onto a phosphor screen, where the emitted photons are coupled into fibers, which are feeding the light to the thermo-electrically cooled CCD sensor. Some specifications are shown in Table 5.1.

5.2.2 *Radiometric Calibration*

The units of intensity measured with an iCCD chip are generally counts over a certain integration time. Counts are generated by changing the capacitor charge state on a CCD by the photo-electric effect from the multiplied photons. Quantitative

Table 5.1 Specifications of PiMax 4 iCCD camera

PiMax-4: 1024F-HB-FG-18-P43	500 kHz	1 MHz	5 MHz
Read noise (e^- RMS)	6.89	7.39	14.49
CCD conversion gain (e^-/count)	1.45	1.49	2.08
Dark charge (e^-/pixel /s)	0.52 (at −25 °)C		
System amplification (counts/pe^-)	15.6 − 454.5		
Minimal gate width (FWHM ns)	2.83		
Insertion delay (ns)	29.8		
Quantum efficiency (%)	50		

knowledge of the collected photon numbers during the integration time can be gained by knowing or measuring all efficiencies from emission to count generation and the sensitivity of the digitizers. Calibrated emission sources with known photon flux in time and direction were used to find the efficiency of the optical system. Changes in configuration, drifts and degradation make this very tedious for a lasting calibration, while cross-calibration with a well known scattering process in the same settings is preferred in most (TS) diagnostic setups [9, 10] and described in Sect. 6.2.

To find out about unexpected deficiencies of the diagnostic setup the efficiencies were cross-checked first against an Ulbricht sphere (USS-600) and additionally with a stabilized, 5 mW laser at 532 nm in two configurations, both in which the notch filter was removed. While calibration with an Ulbricht sphere is commonly employed in spectroscopy, the stabilized laser turned out to be more valuable for gaining information about the optical system. Although the Ulbricht sphere provides a spatially uniform emission profile and a known emission profile, drifts are usually compensated by an integrated photo-diode, which was not available for the measurement. Therefore, only a rough estimate for the overall transmission using the manufacturer-provided emission profile was possible and showed a less than 1% transmission. Since the photon flux of the USS-600 in the wavelength region transmitted by the spectrometer is barely visible, an accurate transmission test was not accomplished, but nevertheless necessary to check the reason(s) for the low transmission. Consequently, the laser light was expanded to the solid angle corresponding to experimental conditions (f = 135 mm, f/2). The power is concentrated at around a narrow spectral bandwidth (0.17 nm) around the laser wavelength rather than being spread over a wide spectral range. This provides enough signal intensity at the various stages of the light path before and inside the spectrometer for power measurements.

Absolute Calibration with Diode Laser

Since the spectral range of the TGS is only 26 nm and the first few nm besides the central wavelength contain the most important signal the usage of a small diode pumped Nd:YAG with stable energy output provides a number of useful information. Besides the absolute performance measurement at 532 nm, the laser is powerful enough to measure the transmission along the optical path through point measurements at every image plane. A diffuse, broadband light source needs to be wavelength filtered and either collimated or viewed from a large distance to imitate the light collection path, while the laser light is already parallel and the <1.8mm spot-size was expanded to match exactly the solid angle of the tele-lens capturing the scattered laser light during Thomson scattering. The laser light inside the TGS focuses on a spot, rather than a spectral line, and its full power can thus be measured by a germanium diode with a diameter of 10 mm. The linearity of this diode was tested beforehand with an array of gray-filters, varying the laser power over four order of magnitude. To exclude general doubts about the specified laser power and iCCD performance, the minimal configuration consists of coupling light via a 2 m fiber directly on to the iCCD camera. The laser (with gray filters) was again cross-checked with a wavelength filtered (few nm bandwidth around 532 nm) USS-600 emission and the quantum efficiency of 50% of the iCCD photo-cathode could be confirmed at 532 nm. The number of photons arriving on the iCCD was directly counted with the method explained in Sect. 6.4 or estimated by the number of counts. The total number of emitted photons N_{Ph} at each wavelength was estimated by emission power and energy per photon:

$$P = \frac{E}{t} = \frac{N_{\mathrm{Ph}}}{t}\frac{hc}{\lambda}$$

$$\frac{N_{\mathrm{Ph}}}{t} = \frac{P\,\lambda}{hc} = \frac{5\,\mathrm{mW} \times 532\,\mathrm{nm}}{1.99 \cdot 10^{-25}\,\mathrm{J\,m}} \cong 1.34 \cdot 10^{16}\,\frac{\mathrm{photons}}{\mathrm{s}}, \tag{5.1}$$

which was then reduced by gating time of a few ns and gray filters to individual photon counts, or by the filter transmission function for the USS-600.

The real transmissions of optical components differ somewhat from the manufacturers specifications and the final transmission is substantially lower than expected. Table 5.2 lists the theoretical values given by manufacturer and measured values of all components, including the laser, as they are passed by the laser light and scattered light along the beam path. Only the Brewster window transmissions were not measured, but the absence of reflected light at the correct polarization hints towards negligible losses. The vacuum windows at PSI-2 are coated over time by a metallic film, especially for experiments with high sputtering yields. Mechanical shutters prevent most depositions, however during the TS measurements the windows are unprotected. A newly developed shutter design is presented in the outlook in Sect. 9.1. After overhauling the laser and replacing all dielectric mirrors to nearly achieve the corresponding theoretical values some deficits of the scattered light collection system

Table 5.2 Efficiency comparison for laser path and collection optics

Component	Specification	Measured
Innolas spitlight 2000	1200 mJ	1100 mJ
Dielectric mirrors: 3x/3x s/p-polarized	$(0.998)^3 \times (0.98)^3 = 0.9356$	$(0.97)^6 = 0.833$
Focusing mirror	0.9	0.70 – 0.8
Vacuum window	0.92	0.50 – 0.92
Lens	~0.9	0.8
30 m fiber coupling and transmission	0.89	0.6
3 holographic gratings and 6 lenses	$(0.75 \times 0.98^2)^3 = 0.37$	$(0.45)^3 - (0.5)^3 = 0.091 - 0.125$
Quantum efficiency iCCD	50	50

remained. The transmission of each grating-lens arrangement was tested separately, hence a "broken" grating could be excluded and a lower than stated transmission in all three components had to be accepted. In total, the (ideal) transmission overall is 1.1% (11.6%), while the transmission of only the light from scattering volume to the iCCD is about 2% (13.8%).

5.2.3 *Timing*

The timing scheme to synchronize Pi-Max 4 spectrometer and V711 fast camera is shown in Fig. 5.4. The reference time base in the first row is determined by the laser using trigger signals of flash-lamp and Pockels cell to trigger fast camera and spectrometer, respectively. Usually, an external signal generator triggers laser and spectrometer for the Pockels cell trigger is too slow to trigger the spectrometer and the flash-lamp trigger too inaccurate.

However, in the setup used at PSI-2 the laser light travels 30 m to the chamber and the scattered light another 30 m in fibers (optical path even longer), while the laser and spectrometer are only a couple meters apart and the spectrometer camera has an input delay of only 30 ns. To trigger both cameras at twice the laser frequency, the laser triggers are fed to a LeCroy crate stack with level shifters (Model 688AL) and double pulse generators (Model 222), which copy and delay an additional pulse to each camera by 50 ms.

The flash-lamp TTL pulse is rising about 215 μs before the laser pulse and thus perfectly suited to trigger the V711 camera well in advance of the laser pulse, indicated by the dashed green line on the left in Fig. 5.4. In burst mode with the typically used settings, the camera captures 100 frames with 2.5 μs exposure time, while an additional delay of 60 ms shifts the captured frames in the fast camera time base to surround the time of the laser pulse. The timing accuracy of the fast camera is stated with up to 20 ns. Additional jitter caused by other timing components of more than 1.25 μs would cause the laser being visible in (on average) multiple frames in the commonly used exposure setting of 2.5 μs. However, the laser containing frame was always found to be stable, hence the overall time-resolution is set by the exposure time of the fast camera.

While the Pockels cell initiates the laser pulse, the corresponding TTL pulse triggers the Pi-Max 4 spectrometer camera after passing through the LeCroy modules with an uncritical delay of roughly 20 ns. An input delay of 218 ns is used with a gate (exposure time) of 12 ns. The exact timing delay of the spectrometer camera versus scattered laser light illumination was fine-tuned by a sequential shift of the delay over 20–30 ns in steps of 1 ns, revealing the time of the highest signal intensity. Using the minimal iCCD gating of 3 ns the average laser pulse shape was scanned and since the laser dump is located 2 m away from the scattering volume the stray-light from the light dump would be visible as a secondary peak 12 ns after the initial signal.

While the synchronization of both cameras to the laser pulse is crucial for the data interpretation, the 50 ms delayed, second trigger pulse (indicated in red) enables

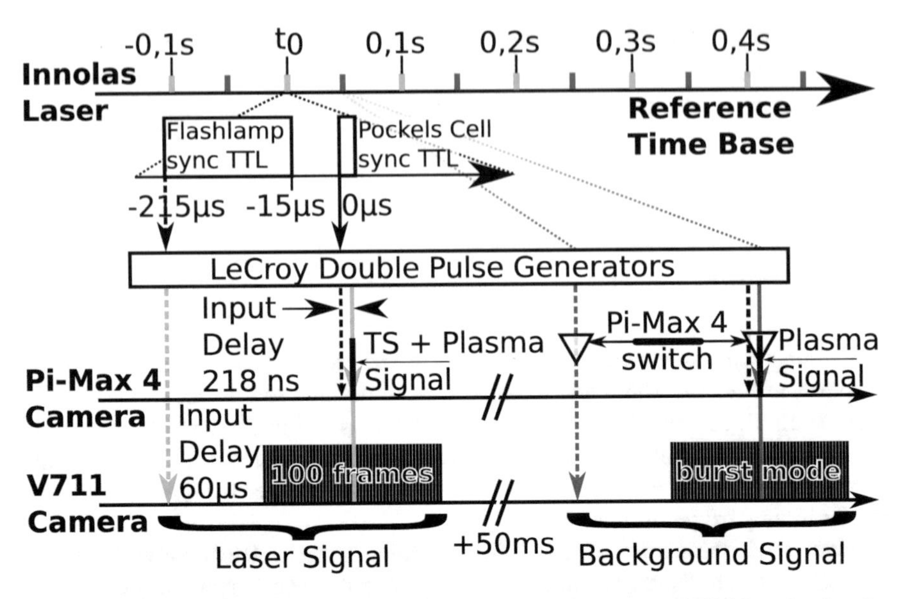

Fig. 5.4 Timing scheme for synchronized TS (Pi-Max 4) and fast camera (V711) based on Innolas laser reference trigger of flash-lamp and Pockels cell

simultaneous measurements of the plasma emission received by the Pi-Max 4 spectrometer camera. Thereby no additional plasma background measurement is needed for background subtraction under the reasonable assumption, that no phase-locked plasma emission fluctuations are present at 10 Hz. Although the timing scheme is more complex by utilizing the laser as the reference time base, the necessity of absolute synchronization of spectrometer and fast camera requires this effort, since intermittent, not regularly and phase-stable occurring, plasma phenomena are the measurement objective. Moreover, counting the TS signal individual to each laser shot and fast camera observation is required and fundamental for employing the photon counting method in Sect. 6.4.

Furthermore, the background plasma emissions received by the spectrometer camera could be correlated to the fast camera signal by an additional 100 frames, especially for brighter transient events, which would then require a variable plasma background subtraction. However, the narrow spectral range of the spectrometer camera signal and the overall low background emissions of the plasma did not necessitate a variable plasma background evaluation thus far and was deactivated by the indicated "Pi-Max 4"-switch.

The synchronized timing scheme is expandable by triggering the second fast camera and variable with the position of the cameras as long as one fast camera and the TS system observe comparable plasma dynamics. Therefore, additional spatial information could be collected in an advanced setup, adding structural knowledge in a particular direction. Furthermore, the V641 camera is spectroscopically valuable if, like for Argon, the neutral and ion transition lines fall into the spectral channels of the color sensor.

5.2.4 Spatial Calibration

The resolution limits of the spectrometer are clearly defined by the set of optical fibers collecting the scattered light, which are arranged as described along with Fig. 5.3. The fibers provide a spatial resolution of roughly 3 mm in their honeycomb arranged entrance slit, but form a line of 1 mm width when back-illuminated through the spectrometer by a red line-laser. To position the focus of the collection optics into the scattering volume and determine the exact resolution a metal plate was mounted on a sample holder and moved by the target manipulator (cf. Fig. 4.1) into the TS section. The plate surface is then bent and rotated parallel to the laser plane.

The fiber projections of the red laser light and the green Nd:YAG adjustment diode laser coaxial to the Innolas main laser are visible as diffuse reflections on the sand-blasted plate shown Fig. 5.5. The larger portion of the image depicts the ideally positioned plate such that the adjustment diode laser spreads grazing over the plate. When correctly aligned and in focus, the red line-laser illuminating the fiber-bundle forms a sharp line of dots across the whole plate parallel to the green reflections. The magnified inlet in Fig. 5.5 shows additional marks each cm for counting the dots per cm and thus inferring the spatial resolution.

Since the collection optics uses a commercial camera lens, the chromatic aberration is negligible and using a red laser eases the calibration by the clear color distinction. Calibrating the fast camera is done in a similar fashion for the simultaneous measurements, while shooting videos at different positions used the Langmuir probe or the known dimensions of the manipulator head as scale reference.

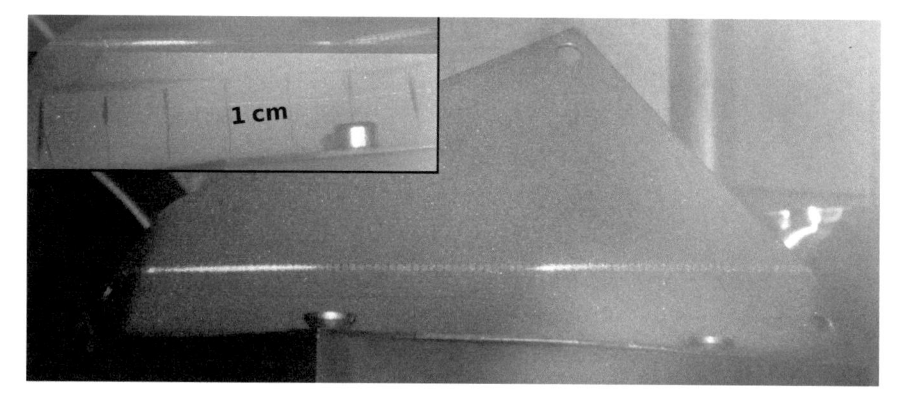

Fig. 5.5 Resolution and calibration aid based on sand-blasted plate at the top of the target manipulator (Fig. 4.1). The inlet shows a magnified image with 1 cm spaced markings on the plate

References

1. Kono A, Nakatani K (2000) Efficient multichannel thomson scattering measurement system for diagnostics of low-temperature plasmas. Rev Sci Instrum 71(7):2716–2721
2. Warner K, Hieftje GM (2002) Thomson scattering from analytical plasmas. Spectrochim Acta Part B AtIc Spectrosc 57(2):201–241
3. Muraoka K, Kono A (2011) Laser thomson scattering for low-temperature plasmas. J Phys D Appl Phys 44(4):043001. (article id. 043001, p. 15)
4. http://www.innolas-laser.com/Products/Lamp-Pumped-Lasers/SpitLight-High-Power.html, 03 2017
5. van der Meiden HJ, Lof AR, van den Berg MA et al (2012) Advanced thomson scattering system for high-flux linear plasma generator. Rev Sci Instrum 83(12):123505
6. Chen H, Zang Q, Han XF et al (2015) Automatic beam alignment system for thomson scattering diagnostic on experimental advanced superconducting tokamak. J Fusion Energy 34(5):1051–1059
7. Strixner D, Email correspondence
8. http://www.princetoninstruments.com/products/PI-MAX4-emICCD, 03 2017
9. Röhr H (1981) Rotational raman scattering of hydrogen and deuterium for calibrating thomson scattering devices. Phys Lett A 81(8):451–453
10. Carbone E, Nijdam S (2015) Thomson scattering on non-equilibrium low density plasmas: principles, practice and challenges. Plasma Phys Control Fusion 57(1):014026. (article id. 014026)

Chapter 6
Data Analysis and Calibration

In this chapter data analysis and calibration procedure are described along with the theoretical description of the laser scattering processes in Sect. 3.1 and experimental setup from Chaps. 4 and 5. The signal processing involves two synchronized diagnostic systems, which are used in special settings and analyzed by newly developed algorithms. Photon counting with extremely short gating times on a 1 Megapixel CCD has recently been demonstrated [1] and allows TS spectra to be time-resolved in software *posteriori*. The signal processing, calibration and operation of the TS system is described first and in more detail. Second, the fast camera data analysis is presented briefly and the processing, including conditional averaging, completes the method chapter.

6.1 Spectral Calibration and Resolution

The scattering of an electromagnetic wave off free electrons in a plasma is a suited method to infer about plasma properties if their interaction is well characterized and the incident electromagnetic wave is not altering the plasma state as estimated in Sect. 3.2. The plasma conditions at PSI-2 allow using the incoherent TS assumptions (Eq. 3.6), hence the scattered Thomson spectra contain information on the electrons only. For the used diagnostic setup the ion feature is indistinguishable from Rayleigh scattering and the central electron spectrum and therefore lost for interpretation. However, the absence of coherent effects creates a direct correspondence of the electron velocity distribution to the measured spectra, folded by the instrumental function.

The spectral calibration was performed with a Neon gas discharge lamp. The TS light collecting lens was focused on the central part of the lamp and a few selected

M. Hubeny, *The Dynamics of Electrons in Linear Plasma Devices and Its Impact on Plasma Surface Interaction*, Springer Theses,
https://doi.org/10.1007/978-3-030-12536-3_6

and strongest lines are compared to several corresponding line transitions from the NIST database [2]. The logarithmic plot of the intensities at the smallest intermediate slit width $d_{\mathrm{Slit}} = 100\,\mu\mathrm{m}$ is shown in the upper part of Fig. 6.1, where many lines are visible and each line is curved along the radial axis on the CCD chip. The curvature is caused by (higher order) image aberrations within the triple grating spectrometer (TGS) and is corrected with a de-skewing routine based on finding the wavelength shift relative to the central spectrum by cross-correlating the radial channels. Then, the spectrum can be fitted against theoretical values of the wavelengths of the lines using only the averaged spectrum in one dimension. Since the curvature is a known effect of high etendue spectrometers, a curved entrance slit can be used to avoid this aberration [3].

Figure 6.1 shows a comparison of Neon spectra in the lower picture at three slit widths d_{Slit}, with the NIST values indicated at the line transition wavelengths. Upon opening the intermediate slit the line width increases as the spectra from the outer fiber rows are imaged on the CCD as well.

A linear dispersion fit of all marked Neon lines compared to the experimental peak position is shown in Fig. 6.2, where the slope results in the wavelength conversion

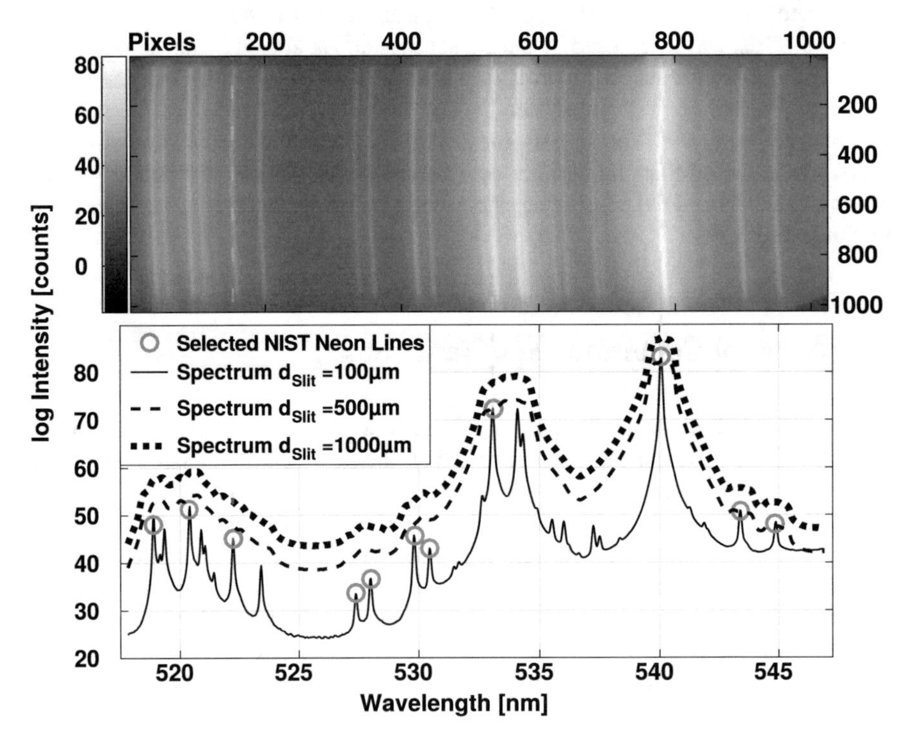

Fig. 6.1 Measured Neon spectrum at $d_{\mathrm{Slit}} = 0.1\,\mathrm{mm}$. The upper picture shows the raw image at full resolution with logarithmic intensity scale. The lower picture compares spectra with different d_{Slit} and the theoretical NIST values

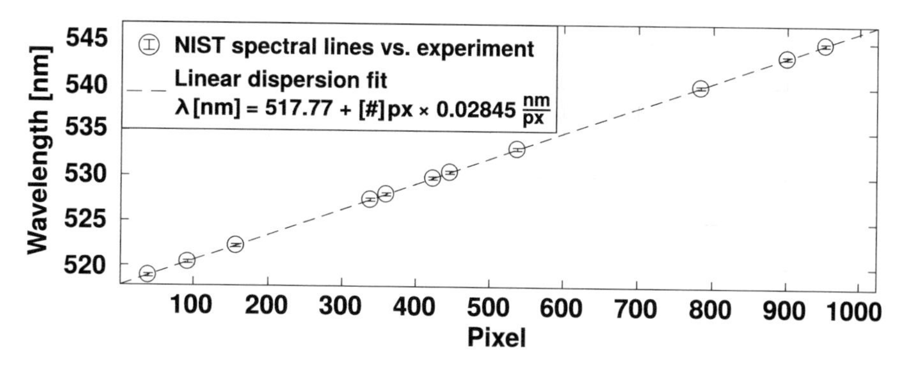

Fig. 6.2 Linear fit to determine spectrometer dispersion

factor of 0.028 nm per pixel. This factor was already used in Fig. 6.1 and the discrepancies between theoretical and measured lines are well below the uncertainty of the instrumental function at $d_{\text{Slit}} = 0.1$ mm. Thus, a linear fit is absolutely sufficient for measurements at $d_{\text{Slit}} = 1$ mm, where individual lines are blurred and no fit can be performed.

An estimate for the instrumental function is done by the width of the Neon line at 540.056 nm from Fig. 6.1, which is the overall strongest and not affected by lines close by. Figure 6.3 shows the normalized and zoomed extract of this line intensity on the left with line widths at half heights (FWHM) for the same three slit widths of Fig. 6.1. Since the fiber rows are each shifted by 275 μm in the "entrance" slit the positioning of the rows is the main contributor to the instrumental function of the TGS. This spectral shift is illustrated in the right part of Fig. 6.3, where each row of fibers corresponds to a substructure in the total perceived instrumental function at a slit width of $d_{\text{Slit}} = 1$ mm. The additional broadening must be accounted for

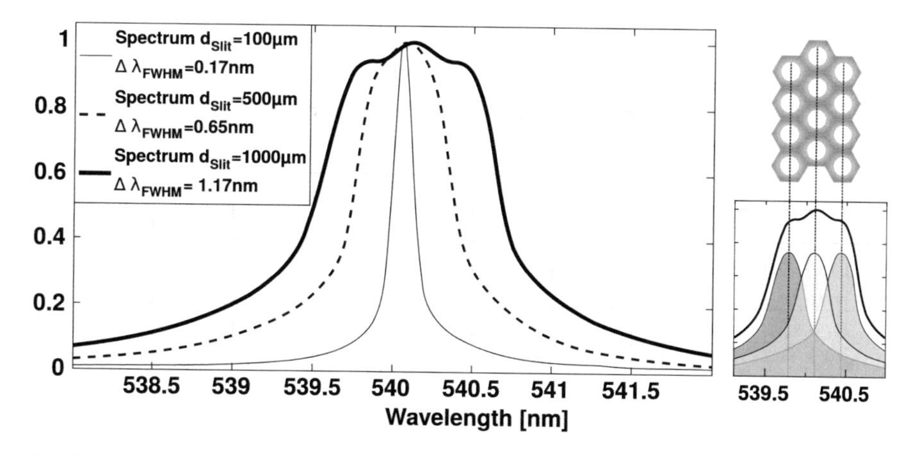

Fig. 6.3 Normalized intensity of brightest Neon line for three different slit widths and their spectral half width estimates

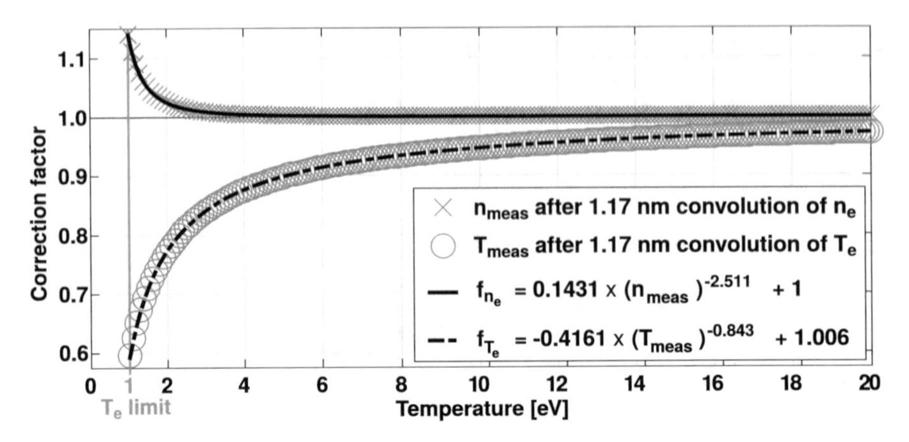

Fig. 6.4 Ratio of density and temperature between test and convoluted Gaussian spectra

and requires a numerical method for deconvolution for each value of d_{Slit}, since the instrumental function is not fitted by a single distribution.

The aberrations causing the curved line shape are also responsible for the small shift of 0.06 nm in central wavelength, as the real slit is imaged skewed at the intermediate slit, thus cutting off parts of the central fiber row at small d_{Slit}. This small shift of the absolute wavelength has no influence on the measurements as it merely adds an offset and only the spectral widths are of importance.

The spectral resolution of 1.17 nm at $d_{\text{Slit}} = 1$ mm causes a significant increase of the measured line width at low temperatures. Therefore, a numerical convolution of an ideal Gaussian distribution and the instrumental function was performed to derive a formula for deconvoluting the measured TS spectra as follows.

For each data point in Fig. 6.4, a Gaussian distribution was folded with the instrumental function and then fitted again with a Gaussian distribution. To mimic the TGS, the spectral part blocked by the commonly used, medium (2 mm) notch filter was removed before fitting. The alterations of area and spectral width are approximated by a numeric equation, defining the detectable, minimal spectral width, i.e. the lowest measurable temperature of 1 eV (red line), which is sufficient for the plasma conditions at PSI-2. Smaller values of d_{Slit} in combination with the 1 mm notch filter could extend the minimal temperature. However, the dramatic loss of photons reduces this option to special cases, in which additional constraints to the setup are applied.

Besides the reduction of the minimal temperature, details of the TS spectrum and a higher spatial resolution are accessible if the intermediate slit is reduced below $d_{\text{Slit}} = 0.25$ mm, allowing only transmission from the central row of fibers in the entrance slit (cf. Fig. 6.4 right). By using the smallest (1 mm) notch filter, only half of the central spectrum is blocked (compared to 2 mm) and some of the photon count reduction is compensated. With a new deconvolution for the reduced instrumental function in this proposed setting, TS spectra with 1 mm radial resolution could be obtained, allow conclusions towards the shape of the EEDF and test for deviations from a Maxwellian distribution.

6.2 Absolute Calibration with Raman Scattering

The absolute calibration of an optical system with a specified light source was used in Sect. 5.2.2 to examine the sensitivity of the system. Raman scattering (RS) is commonly used to calibrate the optical system and has advantages over a specified light source, since the optical setup remains untouched and only the scattering volume is filled with a different source of scattering particles. Furthermore, the RS spectrum occupies a similar wavelength region compared to TS for the plasma parameters used in this work and blocking the central part of the spectrum is also beneficial for analyzing the RS spectrum. The only disadvantage is the lower scattering cross-section ($\sim 10^{-6}$), which is overcome by using pressures in the range of 1–50 mbar and thus a much higher particle density compared to the plasma density. Consequently, RS has been performed in all experiments to calibrate the TS setup to enable absolute plasma density measurements.

Possibly available calibration gases are H_2, D_2, CO_2 or N_2, with the first two considered a safety hazard at pressures in the millibar range. Another consideration is the contamination of the PSI-2 chamber and especially the plasma source, which excludes CO_2. Thus, N_2 is the gas of choice in this work and is commonly used for calibrating other TS setups [4–6].

Since the scattering geometry and the acquisition settings are the same for RS and TS, the scattered powers from Eqs. 3.1 and 3.14, which would contain the system efficiency η_{total} calculated in Sect. 5.2.2, are divided by each other to achieve a laser power, efficiency and solid angle independent measure for the plasma density:

$$\frac{N_{\text{RS}}}{N_{\text{TS}}} = \frac{N_i\, \eta_{\text{total}}\, n_{N_2}\, L_{\text{Det}} \frac{d\sigma_{\text{RS}}}{d\Omega} \Delta\Omega}{N_i\, \eta_{\text{total}}\, n_e\, L_{\text{Det}} \frac{d\sigma_{\text{TS}}}{d\Omega} \Delta\Omega}$$

$$n_e = n_{N_2} \frac{N_{\text{TS}}\, \frac{d\sigma_{\text{RS}}}{d\Omega}}{N_{\text{RS}}\, \frac{d\sigma_{\text{TS}}}{d\Omega}} = n_{N_2} \frac{N_{\text{TS}} \sum \frac{d\sigma_{J\to J'}}{d\Omega}}{N_{\text{RS}}\, r_e^2}$$

$$n_e = n_{N_2} \frac{N_{\text{TS}}}{N_{\text{RS}}} \times C_{\text{Calib}} \tag{6.1}$$

The gas density n_{N_2} is calculated by the ideal gas law $p = n\, k_B T$. The pressure is measured by a Baratron™ gauge in a range of 1–100 mbar, since the capacitive pressure measurement is very precise and gas species independent. The temperature is obtained by a thermocouple touching the vacuum vessel, assuming equal temperatures of gas and surface. Both pressure and temperature accuracies are below 1% and thus negligible compared to the error of the calibration factor C_{Calib} containing the cross-sections of RS and TS. Here, the major error arises from polarizability anisotropy γ in Eq. 3.17, which accounts for an 8% error. The uncertainty estimates for TS and RS photon numbers depend on the signal strength and measurement dura-

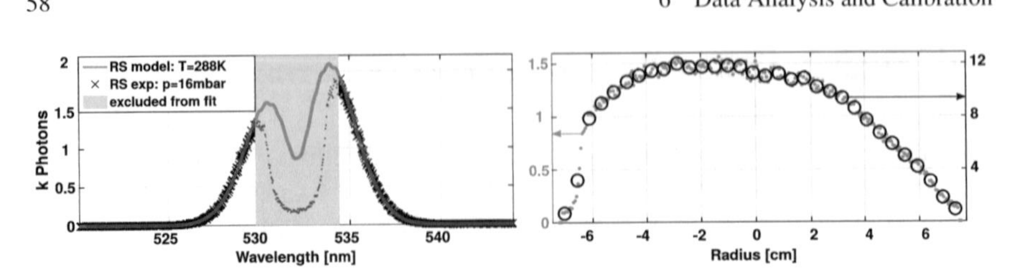

Fig. 6.5 Spectral and radial sum of photons received over 1000 laser shots at 490 mJ. The model equation for the RS is shown in red and the central gray area is blocked by the notch filter and thus not used for the fit

tion, hence it is evaluated for each campaign. To minimize the RS error, variations of pressure and power are performed with a model of the RS spectrum.

A comparison between model and measured RS spectrum can be seen in Fig. 6.5, where on the left side the sum over all radial position is shown. The individual transition lines of the RS spectrum are blurred together (cf. Fig. 3.5), but the shape of the RS is well resembled by the model. The theoretical RS (gray line) is calculated by using Eq. 3.14 with the instrumental line shape from Fig. 6.3. Since there are no deviations from the spectral shape, there is only one factor necessary for the height adjustment, rather than taking the photon number ratios at the individual wavelengths. Instead, the total number of photons received in the scattering process must be calculated by the ratio between area under receiving and blocked spectral region.

On the right side of Fig. 6.5 the spectral sum is shown in native resolution (gray dots) and eightfold binned (circles). The intensity distribution shows an almost flat part and two different decays. At negative radii the decline is abrupt, which is mainly caused by reaching the end of the imaged area on the CCD, thus the outer ~50 Pixel are not used at top and bottom. However, at positive radii the signal intensity drops steadily over several cm. As this shape appeared persistent at various configurations always as the optimal setting for the collections optics (tilt, focus, xyz shift), the origin could be the decreasing laser power density off focus, vignetting in the spectrometer or coupling losses due to a larger tilt of the solid angle towards the edge of the collecting fiber bundle. With these low signal levels, the edge of the useful region is approximately at 6 cm in this setting, depending on the TS signal level. The calibration factor C_{Calib} for absolute density measurements is obtained for each of the binned channels, indicated by circles in Fig. 6.5. In this example the radial resolution is roughly 0.5 cm (15 cm/32 channels).

Variations of pressure and laser power are valuable for estimating the stray-light level, since the real RS signal should depend linearly on these quantities and ideally the signal level extrapolated to origin should be zero. Figure 6.6 shows a laser power variation on the left and a more elaborate variation of laser power and pressure on the right. For all measurements, the sum of all photons is normalized by the shot number and only corrected for the simultaneous background, but not for stray-light. Extending the linearly fitted relation in both pictures to the origin yields an offset

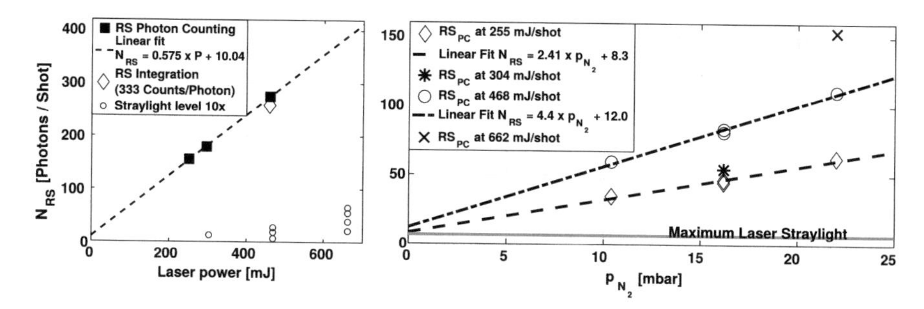

Fig. 6.6 The scaling of the RS signal versus laser power (left) and Nitrogen gas pressure (right) is shown as sum of total photons with three linear fits. Additionally, the converted signal level for CCD integration (left) and an independent stray-light level measurement is indicated (both)

and hence stray-light is present. The linear fit of the laser power dependence in the left graph has an offset of 10 photons per laser shot, which is much higher than the residual background photons received without laser illumination of a maximum 0.05 photons per readout. Measurements of stray-light with and without gas-flow and different times during a day show a strong variation, although they are all far below the signal level with a maximum of 6.5 photons per readout.

Comparing the photon counting method with the CCD integration leads in both cases to the typical spectral and radial shape of the RS. The conversion between counts and photon number is done with the average counts per photons evaluated by the pixel volume estimate (follows in Sect. 6.4), which is 333 counts per photon for the three cases on the linear fit. This results in slightly lower photon counts, while 312 counts per photon would be necessary to exactly match the two points.

The right graph of Fig. 6.6 shows RS measurements at various laser powers and pressures with two laser power dependencies fitted to get the offset at the origin. The stray-light at zero pressure was 8.3 and 12 photons per shot at laser powers of 255 and 468 mJ per shot, respectively, while the actual stray-light measured at $p = 10^{-6}$ mbar was 6 photons per shot. The difference in the offsets at zero pressure in conjunction with an offset at zero laser power points towards a non-linear dependence of the stray-light on the laser power, since the laser independent background is negligible.

The beam quality for a high power laser is ideally independent of the power level, but the internal laser amplification processes depend on power densities and temperature profiles (thermal lensing). The beam quality is usually optimized for the maximum power output [7], which could not be used in the current setup. Therefore, the focusing of the laser beam could explain the difference in offset levels. However, this can be avoided by using the same or closest laser powers during calibration. Furthermore, the spectral shape of the stray-light has distinct features far away from the RS and TS occupied spectral regions, thus the removal of the reproducible, average stray-light level can be achieved.

6.3 Thomson Scattering

The true Thomson scattering signal must be extracted from the measured signal, which includes the unavoidable plasma background and laser stray-light. Raman scattering is another source of unwanted signal in discharges with molecular gases and neutral densities equal or higher than the plasma density. This is especially important in atmospheric plasmas or combustion processes, while at PSI-2 low power Deuterium discharges can be dominated by recombination and thus exert a higher level of neutrals. However, the discharges used for turbulence investigations in this work are highly powered with low neutral densities. The spectral intensity of raw signals collected in two different discharges is shown in Fig. 6.7, where the areas under the curves represent the contributions of TS signal, laser stray-light and the plasma background for Argon (left) and Deuterium (right). The Argon spectrum is clearly affected mostly by stray-light with marginal plasma emissions, while the opposite is visible for the Deuterium case. The summation of laser stray-light and the plasma background signal plotted on top of each other shows that all signal in the region beyond 543 nm is accounted by these two parts. While the plasma background is measured simultaneously and can be subtracted directly, the laser stray-light must be measured before, after and/or in-between plasma operation. Therefore, the total amount of laser stray-light is initially determined by weighting duration and laser power and then scaled to zero the remaining signal beyond 543 nm. Ideally, laser stray-light is minimized to negligible levels, but during plasma operation various drifts in the experimental setup can cause increased stray-light levels, which are then accounted for by this scaling.

The two-dimensional Thomson spectrum after subtracting plasma background and laser stray-light is shown in Fig. 6.8 on the left, while two example spectra are shown on the right. The initial iCCD resolution was reduced to 64 by 64 pixels corresponding to a spatial resolution of 2.3 mm and a spectral resolution of 0.455 nm. With the matching two-dimensional Raman spectrum, the pixel lines of each spectra are paired up and the density is then determined via Eq. 6.1. Temperature and density are already adjusted by the deconvolution.

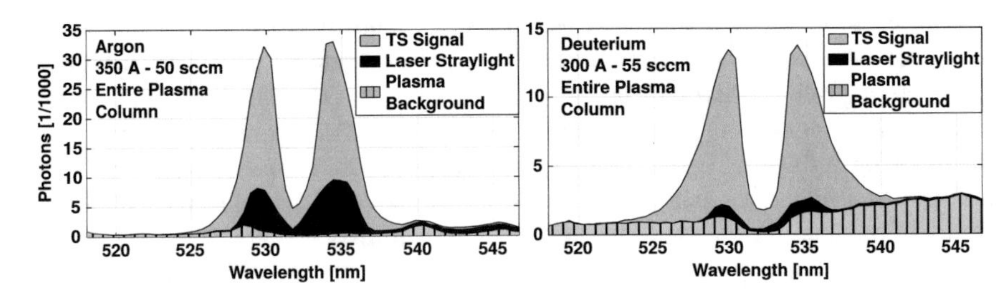

Fig. 6.7 Area under curves shows the radial sum of raw intensities of Thomson signal, laser stray-light and plasma background for Argon and Deuterium discharges, respectively

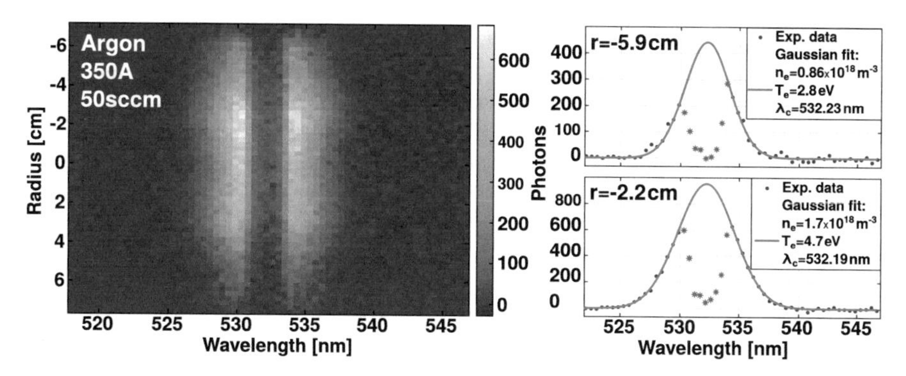

Fig. 6.8 Two-dimensional TS spectrum (left) and example fits and results at two radial positions (right) at a radial resolution of 2.3 mm

The error estimate for the temperature is based on the 95% confidence intervals in a least-squares Gaussian fitting routine with the least-absolute-residuals option (increasing robustness of fit) enabled in MATLAB™. Since for high photon numbers the Poisson distribution used for photon (shot) noise estimates approaches the Gaussian distribution, the standard deviation σ is used for the weighting of the spectrum. To estimate the density error, a combination of fitting error and σ for Raman and Thomson spectra is used. The width of the excluded area in the center is determined by the deviation between RS model and measurement, since the signal levels are higher and less stray-light and no plasma background are present.

6.4 Photon Counting

The integrated signal of an image or spectrum is usually comprised of a number of photons each increasing the count rate on the CCD chip. Besides the regular photons arriving from a intended source, additional photo-electrons are generated over time via electron-hole creation from thermal fluctuations in the intensifier or on the CCD itself, which is called dark charge. Sources of noise are arising from the dark charge but also from the signal amplification in the intensifier and the readout process of the CCD.

For time-averaged spectroscopic measurements, the integration time is chosen long enough to fill the charge capacity of the CCD and using the full dynamic range (e.g. 12/16 bit) of the digitizer, thus avoiding photon noise from Poisson statistics. Although the CCD sensor is generally cooled to e.g. −25 °C by Peltier elements, reducing the dark charge accumulation considerably, integration times of several minutes up to one hour make a dark charge reference measurement necessary. Since the dark charge generation depends on random fluctuations but also on physical properties of each CCD pixel, the mean pattern can be subtracted and only the magnitudes lower photon noise of the dark current itself remains. The digitization (reading) of the charge content of the CCD creates additional noise (thus read-noise) depending on the digitizer frequency but independent of the amount of charge. Therefore, slow digitization of a well filled CCD is preferably used to minimize photon noise, while the high number of collected photons also averages the variance of the individual photon amplification in the intensifier.

As the requirement of this TS setup is observing time-resolved events, the signal collection and reading has to be configured inherently different. As described in Sect. 5.2.3, the signal acquisition must follow twice the laser repetition rate and the TS signal is generated within the 10 ns laser pulse. While there is virtually no dark charge accumulated during this short integration time, the signal from individual photons must be discriminated regarding the amplification distribution in the intensifier and the noise during the digitization. The latter is greatly enhanced compared to time-averaged measurements, since the digitizers are used at or near the fastest settings (cf. read-noise in Table 5.1).

Since the integration method is capturing all photons by design and is commonly applied with TS, some initial measurements were done in order to compare the signal level with the photon counting method, since the algorithm needs to employed properly to not over or underestimate photon numbers, especially when the photon flux is high enough to encounter multiple photons captured in one or adjacent pixels (e.g. two-photon events).

The reference measurements for laser stray-light and plasma background must use the same configuration as the TS signal acquisition. Accumulating the photon numbers from individual shots and plasma emissions over several thousand readouts results in the same spectrum if the photon counting algorithm is employed correctly. Therefore,a detailed description of the algorithm implementation follows.

With typical laser durations of 10 ns the collection of dark current is negligible and only a fraction of the CCD sensor is illuminated - with individual photons. The amount of signal generated by a single photon depends on whether it created a photo electron at the photo cathode of the sensor (Q.E. = 50%) and then the voltage by which this photo electron is accelerated along the dynodes in the photomultiplier stage. Furthermore, the digitizer converts the acquired charge in the CCD depending on readout frequency and gain settings. To verify the values given by the manufacturer a photon flux of only a few photons per readout, distributed over the whole chip was recorded. The read-noise was recorded separately with the intensifier switched off. The comparison between these different settings is shown in Fig. 6.9, where the close distribution around zero represents the read-noise and the long tails on the right side are signal-related. The slowest readout at the 0.5 kHz digitizer setting takes about 2.2 s for a full resolution picture. Thus, even with a switched off intensifier, there is a considerable chance of detecting cosmic rays(CR) during this readout time. The counts produced by CR are much higher compared to regular photons and generate charge in several pixels in a row, thus their distribution has to be separated.

Taking a subset of 37 out of 100 frames without large CR events, the minimal read-noise distribution is drawn with a solid black line in Fig. 6.9, while the gray crosses are the full 100 read-outs. Recording individual photons from a low light source (Ulbricht

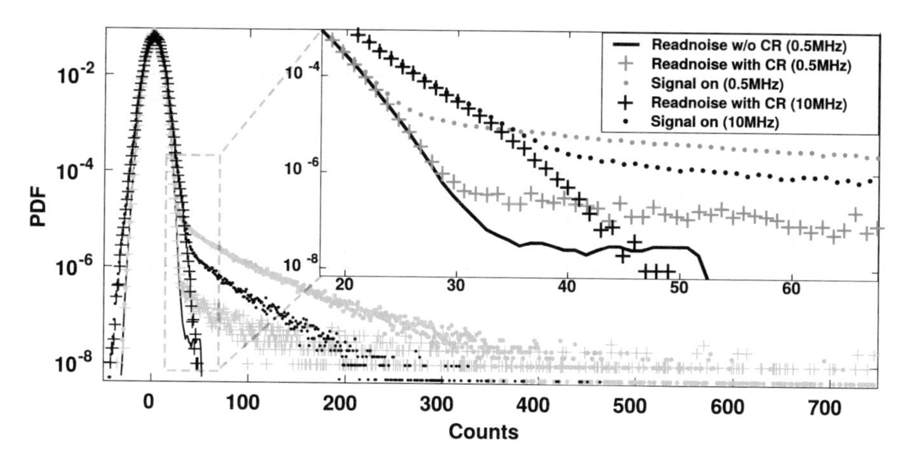

Fig. 6.9 Distribution of read-noise at two digitizer clock speeds and signal levels. The inlet zooms at the threshold, where signal and cosmic rays become significant relative to the read-noise level

sphere USS-600 without lens) results in the dotted gray curve being about a factor 10 larger than the CR. The transition between read-noise and signal is found at 25 and 30 counts for signal-to-read-noise and CR-to-read-noise, respectively. Another transition for a digitizer rate of 10 MHz (2 × 5 MHz) is shown in the main picture and zoomed inlet between black crosses and black dots. The faster readout speed increases the noise in the digitizer and electronic amplification circuits and hence causes a broader read-noise distribution, which is consistent with the values given by the manufacturer. The transition count rate for signal-to(-read)-noise discrimination moves to 35 counts.

Although the same integration time was used for the slow and fast readout case, the CR level is practically non-existent after 50 counts. This means that CR events are mainly captured because of and actually during the long readout time. Furthermore, the slope of the PDF differs for the two readout speeds, pointing towards a different measure of counts per photons, which is already indicated by Table 4.1.

The threshold for the photon detection method in this work is set to 50 counts, which gives a two magnitude margin between read-noise/CR and the actual signal. However, that does not mean every pixel with a count higher than 50 is a single photon. Every intensified photon (i.e. many photons) creates a spatial distribution spreading over several pixels, which is caused by a lower resolution of the intensifier phosphor screen compared to the CCD. Although cross-talk between pixels at these low count rates are negligible, a leakage of amplified electrons in the intensifier channels would result in a similar and thus indistinguishable spatial distribution.

To find the real total counts per photon, the one-dimensional MATLAB™ *findpeaks* routine was applied horizontally (x) and vertically (y) for each pixel line with an intensity threshold of 50 counts. Typically, the horizontal and vertical runs find each several maxima in adjacent lines, while within a line only one clear peak is caused by a photon. The pixel positions of the found peaks are marked and then

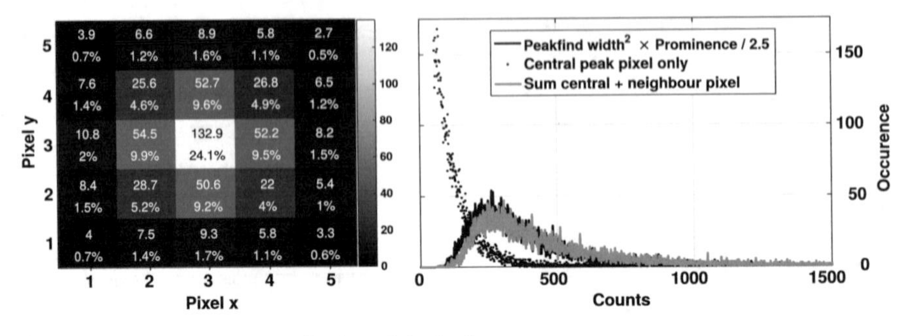

Fig. 6.10 Spatial intensity distribution of single photon

overlayed to identify the central pixel of the photon. Then, the algorithm adds the marked peak positions for the two directions logically (AND), hence only one position will be left as the true center. The resulting spatial distribution of counts for a slow readout is presented in Fig. 6.10 on the left, where one quarter of the counts (133) is in the central pixel and another large fraction (38 and 19% $\simeq$ 300 counts) distributed equally in the adjacent pixels.

The total counts per photons are evaluated by two different methods. The first is simply adding up all the surrounding pixels e.g. from the presented 5 by 5 matrix, resulting in 536 counts per photon. With increasing photon numbers, the probability of catching another close-by photon (within the 5 by 5 matrix) leads to double counting and therefore an overestimation of the counts per photon. A second method uses the so-called peak prominence $p_{p,x/y}$ and half-width $w_{p,x/y}$ information in x and y direction supplied by the *findpeaks* routine to calculate an effective pixel volume $V = w^2\, p/2.5$ with $w = \sqrt{w_{p,x}^2 + w_{p,y}^2}$ and $p = (p_{p,x} + p_{p,y})/2$. The factor 2.5 is to account for the shape of the pixel volume being between a pyramid and two-dimensional Gaussian, since it is strongly limited by the resolution. While the prominence is simply the peak count value, using the "half-widths"-option in the *findpeaks* routine increases the accuracy when two photons are separated by one or two pixels by only taking counts from one side to calculate the half-width. The total counts per photon for this second method results in 384 counts per photon.

The right side of Fig. 6.10 compares the distribution of both counting methods and also displays the central pixel value only. The latter underestimates the counts per photon drastically, while the two other methods agree quite well. The difference in the average counts per photon estimate can be explained by the inherent over- and underestimation of the two counting methods, with the average being 460 counts per photon. This compares excellent with the system amplification given by the manufacturer listing a value of 454.5 counts per photon at maximum gain and slow readout setting. When the counts of a photon surpass 1000, it is assumed that two photons were collected too close to be separated by the algorithm and thus accounted as a two-photon event.

Timing and Binning

Even though CCD sensor readout speeds have become faster, they still are far from CMOS sensors (cf. Table 4.1). Shifting and reading out charges on a CCD chip limits the frame-rates, but provides high quantum efficiency and linearity. The frequency to be matched by the PiMax camera is 10 Hz for the laser and another frame in-between to simultaneously measure the plasma background. The binning of pixels in hardware and the digitization rate define the readout time (seconds to milliseconds) and thus the frame-rate, as the exposure time (ns) is negligible in comparison. Omitting pixels at the edge of the CCD could be used if certain settings are close to the necessary frame-rate, but since roughly 90% of the CCD is needed, the frame-rate benefits are modest.

To achieve a readout frequency of at least 20 Hz the maximum digitization rate (2×5 MHz) had to be used in combination with hardware pixel binning. Ideally, a 2 by 2 pixel binning would be fast enough and not loose information, since each photon occupies more than one pixel anyways. However, binning pixels on the CCD exhibited drastic differences depending on the dimension in which the binning was in effect, since the charge corresponding to each pixel is shifted consecutively down and side-wards to the digitizers. Binning in spectral direction (in the configuration used here) dramatically increased the width of the read-noise distribution far beyond useful at 10 MHz digitization rate. This might be caused by a hardware flaw or due to a line-shift during the charge shift in CCD register, according to the manufacturer. Therefore, the minimal hardware binning allowing a complete CCD readout in less than 50 ms is 1 by 4, which hardly affects the radial resolution (cf. Table 5.2). The usage of a slower digitization rate (2×1 MHz) with 4 by 8 binning also achieves a 50 ms readout. Although the read-noise level is still acceptable, the reduced resolution increases the chance of double photon events by a factor of 64 and possible triple photon events are recorded, which are hard to differentiate. Therefore, the 1 by 4 binning was chosen for all experiments and used with 2×5 MHz digitization rate.

The comparison between signal acquisition with integration and photon counting was first tested with full resolution and homogeneous illumination to verify the manufacturer settings. Then, spectra of plasma discharges and the Neon calibration lamp were tested at different settings. Finally, the real application test was performed on RS spectra (cf. Fig. 6.6) at the standard timing and binning settings as described above. The fast readout reduces the system amplification from 454 down to 317 counts per photon, hence the determined values of roughly 330 counts per photon provide sufficient agreement to conclude the successful operation of the photon counting algorithm. This algorithm and matching the frequency of the laser provides the basis of time-resolved TS for this work, but the equivalence of the signal acquisition would allow other spectroscopic measurements to use this setup to synchronize their spectra to dynamic changes in plasmas or elsewhere. However, besides the timing and readout limitations, the amount of collected data vastly increases or requires an immediate processing.

6.5 Signal Processing

Fluctuating plasma quantities contain important information in the spectral domain and depending on the physical process responsible for the fluctuations, different techniques may be applied to extract and identify them [8]. Regular harmonic oscillations are the starting point for many linearized equations describing plasma dynamics. Thus the Fourier transformation is a basic method to access the frequency domain of the collected data signals to identify dominant oscillations and the general power spectrum. Furthermore, data samples from different positions can be compared in time and frequency domain with techniques explained briefly in the following. Excluded for these techniques are short transients, which are not resembled in the spectral domain, while the underlying plasma processes leading to intermittent transients can start out as fast growing harmonic oscillations.

Fourier Transform

The concept of the Fourier transform (FT) is named after *Joseph Fourier*, who developed this method in the early 1800's in the study on heat conduction. Generally, FT serves as a bridge between the time domain and the frequency domain and led to the development of transforms in other domains. Various methods based on FT give powerful tools to characterize fluctuating signals. There are three types of domain signals, non-periodic, periodic and discrete. The continuous FT is used to decompose an algebraic waveform into an infinite sum of harmonic oscillations. The opposite calculation from the frequencies domain to the time domain is called inverse Fourier transform. Together the pair of FT functions can be written as

$$X(f) = \int_{-\infty}^{+\infty} x(t)\, e^{-2i\pi ft}\, dt \tag{6.2}$$

$$x(t) = \int_{-\infty}^{+\infty} X(f)\, e^{2i\pi ft}\, df\ , \tag{6.3}$$

where $X(f)$ and $x(t)$ are the waveforms in the frequency and time domain, respectively. For digitized data a discrete version of the FT has to be employed, which introduces certain constraints due to the sampling rate and finite intervals. In the discrete FT, the integrals are replaced by a summation, resulting in the following form

$$X(j) = \frac{1}{N}\sum_{k=0}^{N-1} x(k)e^{-i2\pi jk/N},\quad j = 0, 1, 2 \ldots, N-1\ , \tag{6.4}$$

where j and k are the frequency and time index index, respectively. The total number of samples N poses the lower frequency limit, while the sampling rate f_s sets the limit for resolving high frequencies.

Using Eq. 6.4 for the actual computation would require N^2 complex calculations and thus quickly exceed computation capabilities of regular computers. The numbers of discrete FT computations can be drastically reduced by using the fast Fourier transform (FFT) developed by *James Cooley* and *John Tukey* [9].

Auto and Cross-Correlation

Folding a function or discrete data sets with a reference function or data is called cross-correlation and reveals how similar the signals are and if they contain a propagating feature. Auto-correlation is the comparison a data set with itself and tests for repeating patterns. The implementation for discrete data follows

$$R(m) = \sum_{n=-\infty}^{+\infty} X(n) \cdot Y(n-m) , \tag{6.5}$$

where the two signals $X(m) = x(t_m)$ and $Y(m) = y(t_m)$ are discrete data points collected at times $t_m = m\Delta t_s = m/f_s$. For auto-correlation, zero time delay will give the maximum value and then decline, defining the correlation time of the signal(s). In case of a propagating wave the cross-correlation will have a maximum at the time the wave needs to travel between the two positions. With the velocity of a structure or wave, the physical size of a coherent structure can be found by combining velocity and auto-correlation time. The decay of the maximum cross-correlation value with increasing distance can also be used, but needs to be corrected for different fluctuation amplitudes.

Power Spectrum and Coherence

Performing a Fourier transformation on the correlation function is mathematically equivalent to multiplying the individually Fourier transformed signals in the following way

$$P_{XY} = R(f) = X^*(f) \cdot Y(f) , \tag{6.6}$$

where P_{XY} is the complex valued cross-power spectrum of signals X and Y with the asterisk denoting the complex conjugate of the Fourier transformation. For the auto-power spectra $P_{XX}(f)$ and $P_{YY}(f)$ the complex conjugate cancels the phase information and results in a real valued amplitude power spectrum, while a cross-power spectrum contains the phase information $\Delta\Phi_{XY}(f)$ in addition to the amplitudes.

Two oscillations can have the same frequency, but independent causes. When $P_{XY}(f)$ is calculated for different instances, the relative phase $\Delta\Phi_{XY}$ will be randomly distributed if the oscillations are unconnected. Thus, adding up different realizations of the power spectra can be used as a measure of how connected (coherent) each frequency components of the spectrum are, since only constant phase relations at a particular frequency result in a non-zero value. Consequently, the coherence is defined by

$$C_{XY} = \frac{|\langle P_{XY}(f)\rangle|}{\sqrt{\langle P_{XX}(f)\rangle\langle P_{YY}(f)\rangle}}, \tag{6.7}$$

where $\langle ... \rangle$ denotes an average over k similar realizations:

$$\langle P_{XY}(f)\rangle = \frac{1}{k}\sum_{k} |P^k_{XY}(f)| e^{i\Delta\Phi^k_{XY}(f)}. \tag{6.8}$$

The coherence uses the auto-power spectra as a normalization at each frequency and ranges between 0 (incoherent) and 1 (coherent).

6.6 Conditional Average

The final ingredient for turbulence analysis is a technique called conditional averaging (CA), which was used successfully with Langmuir probes [10–15], reflectometry [16] and fast imaging [17, 18]. In contrast to the analysis of harmonic oscillations, singular, strong excursions of fluctuating plasma parameters are better or only characterized by their specific spatiotemporal shape, while spectral features are, especially in broadband turbulence, indistinguishable from the background fluctuations.

Retrieving a subset of data from the whole time series based on a certain condition is particularly useful for dynamics or events, which appear randomly or intermittent, but have a distinct feature easy to be selected by. These events are then averaged disregarding their timing within a time series and common or average properties properly investigated [19]. The selection conditions are somewhat arbitrary and differ within the literature. Mostly, the selection criteria are based on exceeding the background level or 2–3 times the standard deviation σ of a fluctuating parameter.

In the edge and SOL turbulence such an example are intermittent filaments, which show a distinct positive peak in the signal with a sharp increase and slow decay. CA extended to more dimensions in space or different accompanying signals revealed e.g. the potential structure within the filaments. However, CA would also reveal if there is an underlying periodic oscillation for a certain condition to be reached, which by its limited temporal coherence creates a blueprint wavelet, i.e a characteristic spatiotemporal structure. Especially with more-dimensional and synchronized measurements the plasma state can be observed in detail and a sufficient number of events give access to the generation processes, since other fluctuations even out. For example two interacting wave structures, which become unstable only if they are in a certain phase-relation and thereby reach a set condition, will be made visible by CA for the case of the needed phase-relation. In this way complex dynamics are reduced and easier accessible by a physical description.

In the framework of this work the conditions were chosen from different fast camera measurement and the Thomson scattering setup. The brightness of the plasma is imaged by the fast camera at high temporal and spatial resolution independently of Thomson scattering in various configurations to get a more complete picture of the dynamics (possibly 3D). Conditional averaging in an pseudo-randomly fluctuating plasma disentangles the stable from the unstable, intermittency causing, "spectral" components. The skewness created by this intermittency (patterns/waveforms) and the possible representation of brightness to plasma parameters (temperature and density) make conditional averaging in this setup a necessary testing tool to compare fast camera brightness dynamics to Thomson scattering accessible plasma temperature, density and velocity profile. Parameter profiles from subsets of individual TS shots (spectra) are sorted for analysis, corresponding to averaged brightness dynamics, which are extracted by reaching amplitudes expressed in σ. The generation of the filaments and their impact on (enhancing) erosion are the two main important aspects connected to specific shape fluctuations statistics.

Besides the intermittent events also regular oscillations can be triggered by their amplitude maximum. This can be necessary for a correct phase selection, if the coherence time of an oscillation is short against the probing frequency. The conditional averaged dynamics then represent the typical oscillation over the correlation time.

References

1. Teranishi N (2012) Required conditions for photon-counting image sensors. IEEE Trans Electron Devices 59(8):2199–2205
2. Ralchenko Y, Reader J, Kramida A, Team NA (2015) NIST atomic spectra database. (version 5.3 ed.)
3. van der Meiden HJ, Lof AR, van den Berg MA et al (2012) Advanced Thomson scattering system for high-flux linear plasma generator. Rev Sci Instrum 83(12):123505
4. de Regt JM, Engeln RAH, de Groote FPJ, van der Mullen JAM, Schram DC (1995) Thomson scattering experiments on a 100 mhz inductively coupled plasma calibrated by Raman scattering. Rev Sci Instrum 66(5):3228–3233
5. van de Sande MJ, van der Mullen JJAM (2002) Thomson scattering on a low-pressure, inductively-coupled gas discharge lamp. J Phys D: Appl Phys 35(12):1381–1391
6. Carbone E, Nijdam S (2015) Thomson scattering on non-equilibrium low density plasmas: principles, practice and challenges. Plasma Phys Control Fusion 57(1):014026. (article id. 014026)
7. Strixner D Email correspondence
8. Ritz CP, Powers EJ, Rhodes TL et al (1988) Advanced plasma fluctuation analysis techniques and their impact on fusion research (invited). Rev Sci Instrum 59(8):1739–1744
9. Cooley J, Tukey J (1965) An algorithm for machine calculation of complex fourier series. Math Comput 19(90):297
10. Antar GY, Devynck P, Garbet X, Luckhardt SC (2001) Turbulence intermittency and burst properties in tokamak scrape-off layer. Phys Plasmas 8(5):1612–1624
11. Boedo JA, Rudakov DL, Moyer RA et al (2003) Transport by intermittency in the boundary of the DIII-D tokamak. Phys Plasmas 10(5):1670–1677
12. Barni R, Riccardi C, Pierre T et al (2005) Formation of spiral structures and radial convection in the edge region of a magnetized rotating plasma. New J Phys 7(1):225
13. Müller SH, Theiler C, Fasoli A et al (2009) Studies of blob formation, propagation and transport mechanisms in basic experimental plasmas (torpex and csdx). Plasma Phys Control Fusion 51(5):055020. (article id. 055020, 15 pp)
14. Oldenbürger S, Brandt C, Brochard F, Lemoine N, Bonhomme G (2010) Spectroscopic interpretation and velocimetry analysis of fluctuations in a cylindrical plasma recorded by a fast camera. Rev Sci Instrum 81(6):063505–063505-7
15. Liu HQ (2013) Cross-field motion of plasma blob-filaments and related particle flux in an open magnetic field line configuration on quest. J Nucl Mater

16. Vicente J, Conway GD, Manso ME et al (2014) H-mode filament studies with reflectometry in asdex upgrade. Plasma Phys Control Fusion 56(12):125019. (article id. 125019)
17. Fuchert G, Birkenmeier G, Carralero D et al (2014) Blob properties in l- and h-mode from gas-puff imaging in asdex upgrade. Plasma Phys Control Fusion 56(12):125001. (article id. 125001)
18. Thakur SC, Brandt C, Cui L et al (2014) Multi-instability plasma dynamics during the route to fully developed turbulence in a helicon plasma. Plasma Sources Sci Technol 23(4):044006. (article id. 044006)
19. Block D, Teliban I, Greiner F, Piel A (2006) Prospects and limitations of conditional averaging. Phys Scr 122:25–33

Chapter 7
Steady State Plasma Results

In this chapter the results obtained with the Thomson scattering diagnostic setup are presented and compared to the Langmuir probe results. After examining the general stability of the signal acquisition and discussing options to account for drifts in the setup, the equilibrium profiles of plasma parameters obtained by Thomson scattering are shown. Four different working gases with several discharge parameters each were measured and presented along Langmuir probe results as the standard diagnostic.

7.1 System Stability

Calibration of and measurements with a Thomson scattering diagnostic require stable conditions for the beam path and collection optics over the course of several hours. Apparent changes of measured quantities occur either as a result of drifts in the diagnostic itself or stem from alterations of the plasma. Differentiating and accounting for these changes extends the integration time and thus the accuracy of the measurements, while drifting plasma conditions are of general interest for the operation of PSI-2. Before such plasma discharge changes can be evaluated it is essential to ensure the stability for key aspects of the diagnostic setup, hence the options to correct for low variations are presented first. For this purpose, the photon counting method provides invaluable information during a set of measurements, since a variable number of laser shots can be combined and analyzed. This further helps dissecting the contributions between possible changes of the diagnostic setup from plasma conditions.

The first element of signal stability is represented by the variation of the laser pulse energy from the laser itself towards the scattering volume. To exclude influences of laser output energy, beam path and its transmission losses (cf. Sect. 5.2.2), the pulse

M. Hubeny, *The Dynamics of Electrons in Linear Plasma Devices and Its Impact on Plasma Surface Interaction*, Springer Theses,
https://doi.org/10.1007/978-3-030-12536-3_7

energy of several hundred laser shots is measured by a power meter at the position of the (removed) beam dump. The measured laser power is scanned in three to four steps from low ($\sim$200 mJ/pulse) to high ($\sim$600 mJ/pulse) power, with each step settling at a standard deviation of the energy measurements of 2% down to 1%, respectively. These scans are performed the first time after positioning and calibrating the beam path and then at least a second time at the end of the measurement campaign. The laser pulse energy at the highest power is hence known with an accuracy of 1%, while differences between first and last energy measurements are assumed to arise gradually over time if no indications of an abrupt change are apparent.

Therefore, surveillance cameras monitor the laser beam position on the mirrors constantly and help document drifts during the day by comparing images similar to Fig. 5.1. As long as the beam path drifts within the boundaries of all mirrors and baffles, no changes in pulse energy are expected, while mirror surfaces can slowly degrade over time and subsequently reduce their reflectivity. However, these degradation takes place over much longer time scales than one experimental day.

The next element of the diagnostic stability includes the signal acquisition from the scattering volume to the spectrometer. Besides the sensitivity to a given amount of scattered photons, defined by the calibration, the stray-light levels and the collection optics must be characterized. As mentioned in Sect. 6.2, Raman scattering is performed on Nitrogen with the plasma source at room temperature and thus ideally right before the plasma operation of PSI-2 is started. Variations of laser power and pressure gives additional information about stray-light levels as explained before (Fig. 6.6). Usually, before the plasma operation only a power scan is performed for time reasons, but after plasma operation laser power and Nitrogen pressure are varied. Similarly, the stray-light level is determined for different laser powers after or before Nitrogen is present in the vacuum chamber, but also in between plasma operation with and without the working gas.

Since the Nitrogen gas density and laser power are both measured with high accuracy ($\leq$1%), the further stability of the diagnostic system is defined by the laser beam positioning within the scattering volume. The RS signal level corrected for variations in stray-light, laser power and gas density resembles the sensitivity and thus the absolute density calibration, according to Eq. 6.1. Therefore, a changing sensitivity is primarily connected to the optical collection path including the vacuum window transmission. Vibrations of the plasma chamber and temperature variations induce slow drifts in the optical alignment, while the unprotected vacuum window between plasma and collection optics is prone to metallic deposits, slowly lowering the transmission over time. Unlike for the laser power measurements, small displacements of the beam path, which are visible with the surveillance cameras, reduce the sensitivity significantly (up to 60%) in most measurements. Due to the imaged area with a width of roughly 1 mm and a laser focus of ideally the same dimensions, small shifts could even cause a loss of signal, but the actually achieved laser focus of roughly 2 mm (cf. Sect. 5.1 and Fig. 5.2) allowed for more clearance. However, with properly accounting for the gradual changes in sensitivity the uncertainty of the calibration is not necessarily the full extent of the signal loss. An important consideration are changes in the sensitivity profile, represented by the RS spectrum. Figure 7.1 com-

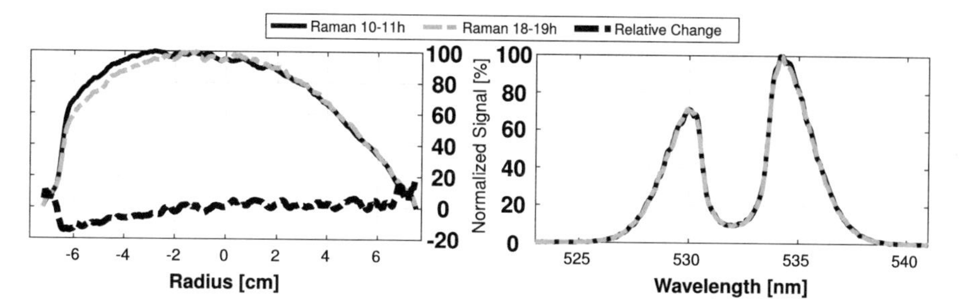

Fig. 7.1 Normalized spatial and spectral average Raman scattering signal for measurements before and after plasma operation with relative change for the spatial average

pares the normalized spatial (left) and spectral (right) RS signals before and after the plasma operation to evaluate the extent of relative profile changes. By cooling the chamber and thus the Nitrogen to 15 °C the spectral shape is unchanged, while the spatial distribution of the RS signal exhibits a stronger decay towards the negative edge. This amounts to a relative change of up to 15% at the edge, which has to be accounted for in the same way as the overall calibration factor is adapted. This will be demonstrated by looking at the signal evolution of a whole campaign day, but first the signal stability is detailed on the scale of minutes.

The necessary requirement for short term stability is given by the low variation of laser power and negligible alignment drifts on short time scales (seconds to minutes). This last level of the stability analysis is based on the photon count rate for each frame exposure as received by the spectrometer camera. An example of all essential quantities is compiled in Fig. 7.2, where the Raman scattering signal (a) is shown on the left for two different laser pulse energies with the almost negligible chamber background. Since the photon count scale is logarithmic, a zero count rate for background signals is included on an extra line. Figure 7.2b shows the laser background without Nitrogen in the chamber and (c) the Thomson signal with the plasma background on the right. Besides the individual frames, marked as dots, time-averaged (25 s) values are indicated by squares. Figure 7.2d shows the signal distribution as a PDF for each of measured quantities, except the chamber background. While the plasma background follows a Poisson distribution, the TS and RS signal are well approximated by a Gaussian function for which the standard deviation represents the statistical error.

The raw TS signal and the plasma background are roughly constant in time, while the laser background increases slightly. For time-averaged TS measurements, the required integration is in the range of several minutes (depending on plasma density, temperature, laser power, etc.), for which averaging is clearly valid. As described in Sect. 6.3, the raw TS signal is comprised of TS signal, laser and plasma background, which are separated by almost one order of magnitude, but also spectrally. As already visible in Fig. 7.2b, the laser background is drifting, but the spectral separation allows an estimate of the laser background evolution throughout the TS measurements. Thus the laser background removal can be more accurate for the analysis of shorter

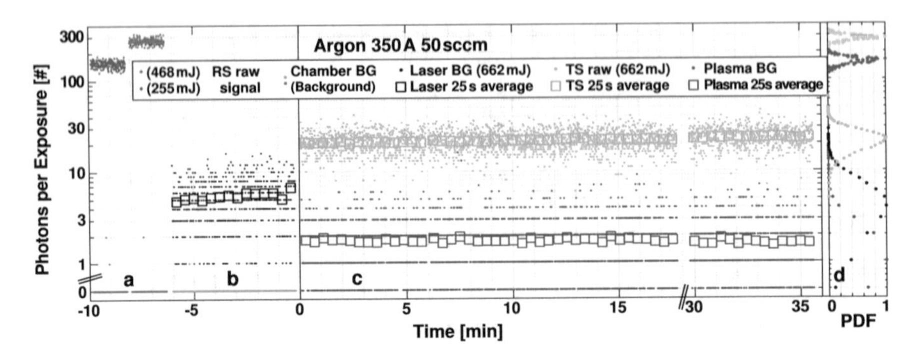

Fig. 7.2 Photons per exposure with or without laser for **a** Raman scattering calibration, **b** laser background, **c** extract of 35 min Argon plasma measurement and **d** show the PDFs of all signals. Note the logarithmic scale with the cut-off below one photon per exposure and an added line for zero counts

samples and give an impression of the drift in between dedicated laser background measurements.

The stability during longer measurements is now evaluated by returning to the time scale of a full (measurement) day. Figure 7.3 shows an overview over all measured and some derived quantities in the upper picture and the average electron temperature and density in the discharge of each collection cycle (2000–7000 laser shots) in the lower picture. The signal acquired over a cycle is summed up and normalized to the used laser pulse energy and pressure (only for Raman scattering).

The comparison between signal levels and actual plasma parameters during the day from the analysis in Fig. 7.3 show a indeed a drastic change between Raman signals before and after the operation of PSI-2, indicating a loss of two-thirds of the signal. An estimate of the laser background evolution for each collection cycle is marked by the laser factor, extending the use of the dedicated laser background measurements. Langmuir probe measurements before and after TS, using identical discharge parameters, give additional information about sensitivity changes. The Raman scale factor estimates the evolution of the calibration factor during the five Argon discharge settings used.

To attribute the signal loss during the course of the eight hours, the changes of measured density within the same discharge parameters are combined and compared. In the first plasma discharge setting with 250 A discharge current and 50 sccm gas flow the Thomson signal level only decreased by about 20% although the average density steadily declined by overall 30%, because the laser background increased as well. Another drastic drop can be seen during the third discharge setting, where the signal almost halfed, while the laser background stays constant. However, the plasma background drops by 30% as well and while this is not straightforwardly translated into decreasing plasma density, a change of the plasma condition must be considered in addition to signal loss. This is contradicted by the Langmuir probe results, which

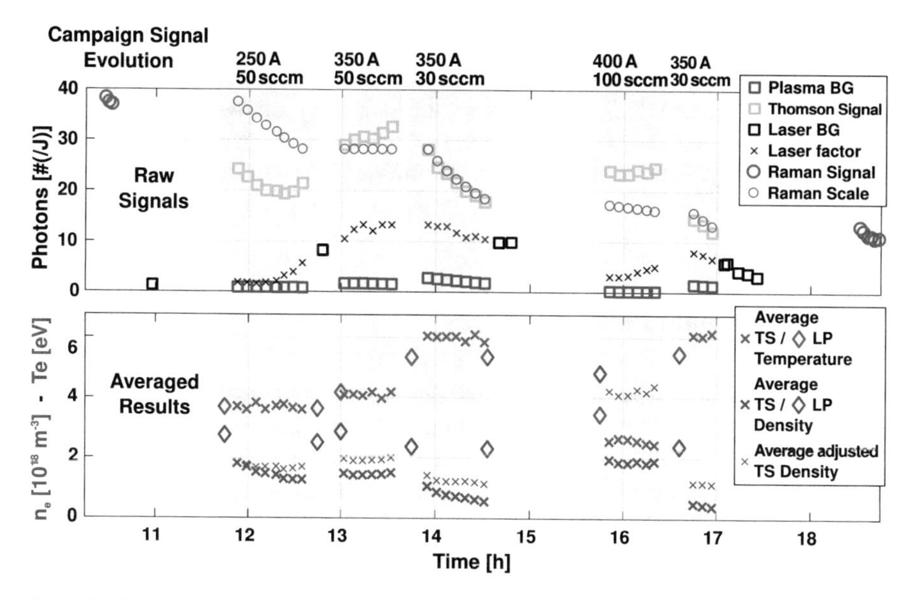

Fig. 7.3 Signal evolution over several plasma conditions with Argon. Normalized (explanation in text) raw signal evolution of Raman and Thomson scattering, plasma background and stray-light level are shown in the upper picture. Below, average plasma parameters derived from TS are compared to Langmuir probe results

remain stable. When comparing the identical discharge parameters of third and fifth setting, another 30% of decline is visible.

In addition, these three contributions explain the overall loss of the signal intensity and are likely caused by a shift of the optical alignment caused by the plasma startup and changing plasma conditions. Especially the third setting used a high thermal and highest power coupling of 30 kW to a low gas feed. Here, a decline of the overall transmission would bring in line LP and TS signal observations. While the transmissivity of the vacuum windows usually does not change this fast, the used discharge settings (especially the third) generated Argon plasma with very high temperatures and hence make higher rates of deposits conceivable. In turn, the conditions for the second and fourth discharge setting are more stable and density calibration factor can be extrapolated without a 60% uncertainty, but rather within the 8% error found in Sect. 6.2.

Since optical adjustment and absolute density calibration with Raman scattering before and after plasma operation are a boundary condition operating PSI-2, the only additional quantitative evaluation about the state of optical beam and collection path during plasma operation is Rayleigh scattering. While this extra information would have been useful on this particular day with frequently changing plasma conditions and thus increased tendency to induce drifts in the optical paths, the Argon pressure, reached with a gas flow of 100 sccm, was $6 \times 10^{-5} - 2 \times 10^{-3}$ mbar. This rather large range of pressures was measured by full range gauges at different chamber stages and is caused by the distribution of vacuum pumps and the large uncertainty of the pressure gauges (∼15%), while none of the more precise Baratron™ gauges is suited for this pressure range. Furthermore, the temporary removal of the notch filter in the spectrometer to collect the Rayleigh signal is an additional source of error and requires generally higher signal levels as the unfiltered laser background is collected as well.

7.2 Steady State Plasma Profiles

Thomson scattering measurements in low density plasmas requires the integration of the scattered laser light over several hundred laser shots to ensure a sufficient signal to noise ratio (SNR). The resulting plasma parameters are thus averages over the measurement time, provided that parameter changes occur uncorrelated to the probing laser frequency of 10 Hz. The required time for a given SNR depends primarily on the density of the plasma for the correlation between density and signal, but also on the temperature and plasma light since a broader spectral distribution of the Thomson spectrum lowers the SNR at each wavelength for a given density. This is especially important as the main objectives are rather hot deuterium plasmas, which tend to have lower density compared to the noble gases.

The standard diagnostic on PSI-2 is the Langmuir probe installed vertically on port F2.2 (cf. Fig. 4.1) with about 30 cm axial distance to the TS section. The probe is operated in single and double probe configuration, thus the plasma conditions during Thomson scattering experiments are also routinely measured by the Langmuir probe, if available. A measurement and subsequent analysis are conveniently completed within one minute and allow direct feedback, since the flux of particles towards the target is needed to estimate the total fluence and hence duration of the plasma exposure. The plasma parameters obtained by the probe are measured from the edge to the center and assumed to be azimuthally symmetric. The routine for probe data analysis assumes the charge state of the working gases to be unity, while ion mass spectrometry results in Argon discharges indicate a fraction of up to 50% doubly ionized Argon ions [1]. To cross-check the results of Thomson scattering and Langmuir probe measurements, the latter are routinely performed before and/or after Thomson measurements and included for most measurements in the following figures. For better visibility only selected Langmuir probe profiles are shown each with dashed lines or marks and same color as the TS measurement conditions, while the error bars are generally in the same range as the TS measurements. The axial and azimuthal positions of the profiles obtained by the probe are not coinciding, therefor the profiles measured at the Langmuir probe position are 30 cm closer to the source and at a roughly 10% higher axial magnetic field. This could lead to reduced cross-field transport losses thus steeper profiles. The comparison between both diagnostics is discussed after presenting profiles for all tested gases.

7.2.1 Argon Discharges

Plasma discharges with the working gas Argon are widely used in many laboratory plasma with various plasma sources for a number of reasons. As a noble gas, no chemistry is introduced in plasma wall interaction studies while the low ionization energy (compared to Helium and Neon) result in comparably high plasma densities and moderate electron temperatures. The first tests of the TS diagnostic were conducted in Argon discharges with high current and high gas feed to achieve the highest SNR to compensate for any initial shortcomings in the setup. Additionally, the Argon plasma light emissions around the laser wavelength are almost negligible.

The first profiles of density, temperature, velocity and pressure presented in Fig. 7.4 are constructed by all the measurements in the second discharge setting presented in Fig. 7.3, but with retaining the spatial information of the scattered light. A total of 21000 laser shots with an average energy of 660 mJ are collected over the course of 35 min and since the photon counting method was used, another 21000 CCD images are acquiring the plasma background simultaneously. From the initial 1024 $\times$ 256 pixel resolution of the CCD the spectral binning is usually 16 to retain 0.45 nm resolution, while the radial resolution is chosen variably depending on the signal intensity. Two profiles are shown for the highest useful resolution of 2.3 mm slightly below the physical resolution (3 mm) and a fourfold lower resolution to reduce error bars from the Gaussian fitting.

Electron temperature and density show their peak values at the projected radius of the plasma source at $\approx \pm 2 - 3$ cm. The density profile is slightly asymmetric with a small dip in the central density, while the temperature profile has a more pronounced hollow profile. This leads to a hollow pressure profile peaking at 2 and 1.5 Pa. The

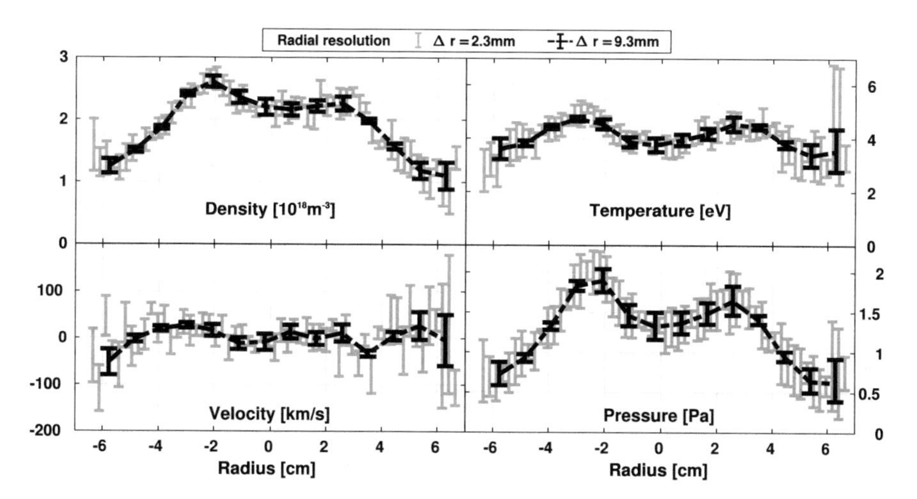

Fig. 7.4 Comparison of two different spatial resolutions for the 350 A 50 sccm Argon discharge profiles of electron density, temperature, velocity and pressure

central wavelength shift of the Thomson spectra are expressed as electron velocities and show a large spread at high velocities in positive and negative direction for the higher resolution profile and a velocity profile within ±50 km s^{-1} for the reduced resolution.

Averaging over two Thomson spectra with identical temperatures but different velocity leads to a higher temperature prediction due to the missing central part of the distribution, especially if the temperature is low. Although the spectral broadening of the spectrometer is not expected to randomly shift the central wavelength, an instrumental resolution of 1 nm corresponds to a shift of 210 km s^{-1}, which calls for caution in interpreting the velocity profile.

The velocity components corresponding to the wavelength shift stem from the radial and azimuthal movement of the electrons during the scattering process and can not be differentiated in this setup. However, the azimuthal velocity is likely to dominate the spectral shift, since the radial movement is restricted by the magnetic field. The rotation frequencies under the assumption of a purely azimuthal velocity reach up to 100 kHz in the left pressure gradient region, but the frequency is only stable over a small radial distance.

The mean electron velocity also corresponds to current densities of up to $\vec{j} = q\, n_e\, v_e = 1 \frac{\text{A}}{\text{cm}^2}$ at r = −3 cm. Although the exposure chamber of PSI-2 is considered current-free (without target bias), responsible for the measured current could be local compensation currents induced by drifts, ensuring the current continuity ($\vec{j}_\perp + \vec{j}_\parallel = 0$) through the axial ($\vec{j}_\parallel$) component, which is not recorded by TS.

A collection of density and temperature profiles for all discharge parameters corresponding to Fig. 7.3 is shown in Fig. 7.5, with the corrected density profiles and only two Langmuir probe profiles for better visibility. The gas-feed has a clear effect on increasing the density, while decreasing the temperature. Depending on the parameters, an almost flat density and temperature profile (blue curve) is obtained, while

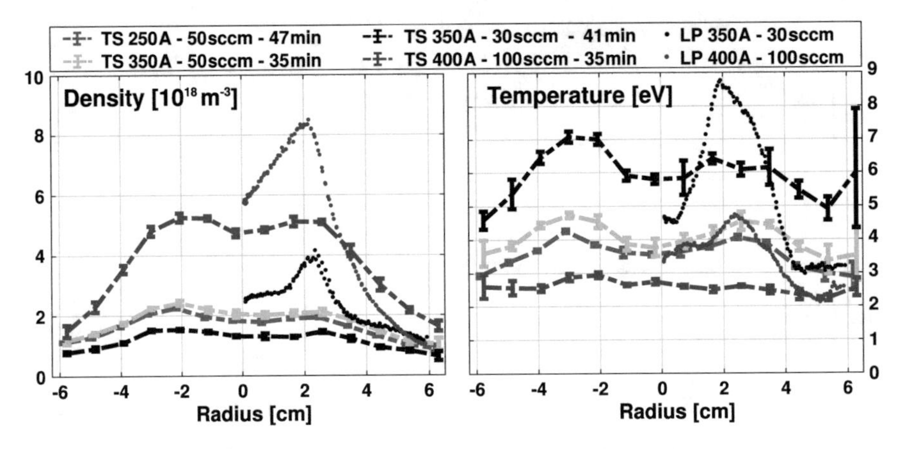

Fig. 7.5 TS density and temperature profiles of all Argon discharge settings compared to two Langmuir probe results

low gas-feed with higher discharge current leads to a slightly hollow temperature profile peaking at 7 eV. While the Langmuir probe measurements share the trend, both profiles of both discharges exhibit much stronger gradients inside and outside the peak position, for which possible reasons are discussed later.

7.2.2 *Neon and Helium Discharges*

Plasma discharges with Neon and Helium are usually used in conjunction with Deuterium to mimic the plasma wall interactions in the SOL with a more realistic gas mixture [2, 3]. Furthermore, their intermediate atomic masses between Argon and Deuterium are convenient to investigate the mass dependence of plasma dynamics or plasma surface interaction, e.g. sputtering thresholds. The increased ionization energy (compared to Argon) and absence of molecular dynamics lead to higher plasma (electron) temperatures and serve as an extension of the parameter space comparing Thomson scattering results to Langmuir probes.

Profiles of electron density and temperature in Neon discharges are shown in Fig. 7.6, where each of the two different discharge settings is compared to Langmuir probe measurements. Similar to the Argon cases, the measurements agree qualitatively. The density drop towards the plasma center is less pronounced with both diagnostics, while the hollow temperature profile in the Thomson profiles is broader. The quantitative differences are strongest around the maximum gradient regions of both profiles. The combination of high electron density and temperature lead to the highest pressure of all investigated discharge with 5–7 Pa in the plasma center.

Similar results are observed for Helium discharges presented in Fig. 7.7, where two Helium discharges are compared with one Langmuir probe measurement. As

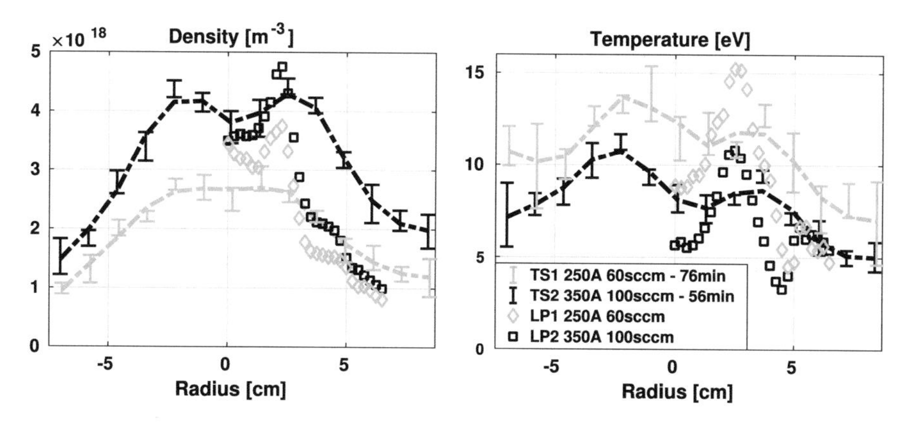

Fig. 7.6 TS density and temperature profiles of two different Neon discharges. Each is accompanied with the corresponding Langmuir probe profiles

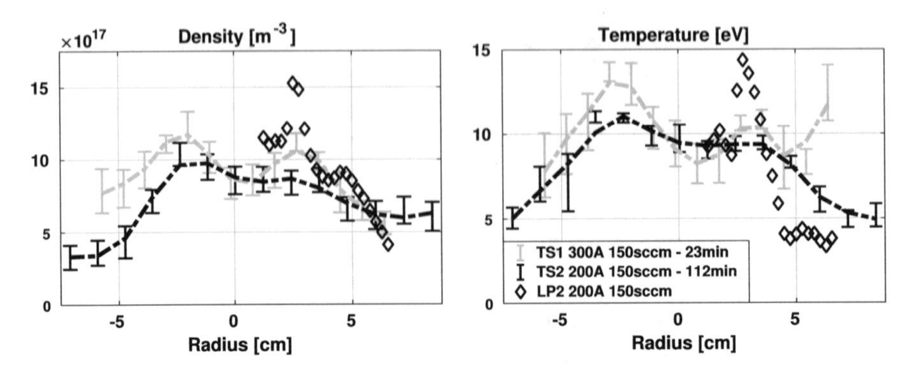

Fig. 7.7 TS density and temperature profiles of Helium discharges on two different days compared with Langmuir probe measurements

for the Neon cases, slightly more pronounced hollow profile shape of density and temperature is measured at higher discharge current.

7.2.3 Deuterium Discharges

Most of the measurements for this work were performed in Deuterium at high input powers and low gas feed. Naturally, Deuterium discharges are most suited to study plasma dynamics and PWI in the fusion context, while Tritium as the other main fuel component is tough to handle for its radioactivity. Compared to the previously presented gas species, the low ion mass of Deuterium leads to a high sputtering threshold on a tungsten surface and thus a strong temperature dependence in the sputtering yield. Therefore, the accurate profiles of the plasma parameters are crucial, especially for high performance plasma with fast dynamics.

An overview and comparison of several experimental days is shown in Fig. 7.8, where density (left) and temperature (right) profiles of Thomson scattering measurements are compared to Langmuir probe measurements. The latter are only shown for two cases for better visibility. The density profile is slightly hollow and peaked at $r \approx \pm 2\,\text{cm}$ with a maximum density around $7 - 11 \times 10^{17}\,\text{m}^{-3}$. The variations of the errors are caused by the measurement duration but also the conditions during each experimental day. Only the black and red curve exhibit errors small enough to verify the hollow profile with a slight asymmetry. Similarly, the temperature profile shows an even stronger asymmetry, or even the absence of the smaller temperature peak on the positive radii-side. The peak temperature ranges between 7 and 10 eV at $r \approx \pm 2.5 - 3\,\text{cm}$. The usually assumed symmetry of the density and temperature profile is contradicted only slightly, but since both parameters are higher on the negative side, the differences add up in the plasma pressure. The comparison with Langmuir probe measurements shows again a similar range with a much more pronounced gradient in the source region.

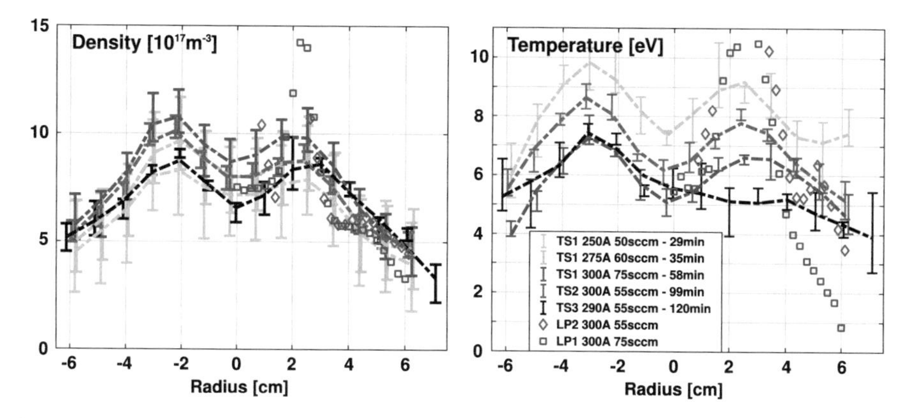

Fig. 7.8 TS density and temperature profiles of Deuterium discharges on different days. On the first day three different discharges are presented with shorter accumulation times (TS1). The corresponding Langmuir probe profiles are same-colored and only presented for two cases for better visibility

7.2.4 Discussion

The measured equilibrium electron density and temperature profiles are found to be in reasonable agreement with current and previous measurements by other plasma diagnostics [4–8]. It was found that the general assumption of an azimuthal symmetry in these profiles is not always given, and depends on the gas species and discharge conditions. However, since mostly high performance discharges were tested and not a complete survey, no general trends can be concluded.

The differences between the Langmuir probe measurements and TS are clearly visible and never coinciding completely, while the general shape is mostly similar with peaking temperature and density profiles at $\approx \pm 2 - 3$ cm. Much higher gradients of Langmuir temperature and density profiles lead to more pronounced and mostly hollow profiles, while trends of increasing or decreasing temperature or densities in changing discharge parameters of the same gas are matched. However, the absolute figures differ by a variable factor of two above and below, hence statistical errors do not sufficiently explain the differences. Furthermore, the deviations are not consistent for different discharges neither in the temperature nor density profile. This indicates a non-trivial or multiple reasons influencing either measurement method, while simply a different prefactor in either theory will not align the measurements. Furthermore, Langmuir measurements at the highest input powers ($\sim$30 kW) were complicated by I-V-curves, which could not be analyzed in or beyond the pressure maximum, due to limitations in software or hardware, e.g. probe emission at high power loads on the probe in the center of the discharge.

One of the most common issues is probably the anisotropy caused by the magnetic field which make Langmuir probe measurements difficult to interpret quantitatively, especially for absolute density determination since the effective collecting area is not well defined, while TS measurements are unaffected by the magnetic field as long as the laser frequency is well above the electron cyclotron frequency and the spectral broadening is determined by the instrumental profile of the spectrometer. Another reason for discrepancies in Argon and Neon discharges with relatively high power input is the presence of double charged ions [9], which lowers the effective mass leading to density overestimation and introduces changes in the exponential increasing part of the I-V-curve. Especially for Argon the fraction of double charged ions was reported with up to up to 50% for discharges with high input power.

Generally, Langmuir probes were found to *change the PSI-2 plasma globally* when measuring profiles with two probes at the same time. The probe acts as an additional sink, reducing the density considerably, while temperatures remained unchanged [5]. Fast camera measurements confirmed changing plasma emissions and their oscillation frequency during the probe measurements. Therefore, part of the deviations in the plasma profile could be induced by the probe itself. The perturbation would then especially affect the regions the plasma must diffuse to, while the flux projected directly from the source to the probe would be less influenced, leading to the observed steeper gradients.

The investigated Deuterium discharges are specifically selected to exhibit strong turbulence and thus fast dynamics in the plasma profiles. Both diagnostics perform temporal averaging over time-scales longer than the gyration duration of both species or other dynamic changes, while the measurement duration itself is 10 ns for the laser shot versus 20 ms for acquiring an I-V-curve. Therefore, changes of plasma parameters during this acquisition time lead to a flattening of the I-V-curve, especially if the plasma potential changes, which do not necessarily correspond to the average density or temperature of the fluctuation.

Lastly, the probe is situated about 30 cm axially upstream (F2.1) and measures only one side of the profile vertically, whereas the TS profiles are inclined 55° from the vertical axis (F3.4). Theoretically, this may only account for part of the deviation, since azimuthal and short axial variations of plasma parameters are expected to be small [10], due to the fast axial transport. However, comparable measurements in the source region in PSI-1 and with a segmented neutralizer plate at the dump position in PSI-2 showed stronger variations of the profile shape [11, 12]. The influence of the flux expansion just upstream of the Langmuir probe position and the slightly higher axial magnetic field could explain part of the profile shape difference present in most measurements. If the high input powers increase the radial transport in the target chamber, these effects should be especially present in the investigated discharge settings. The number of possible uncertainties for Langmuir probe measurements, especially in the discharge settings used in this work, and the perturbation of the plasma by the probe itself result in a demand for additional measurements under these conditions, if used for sample exposures with a need of accurate knowledge of the plasma parameters.

Although the radial resolution of the Langmuir probe is much higher then TS, its 3 mm spatial accuracy should still be able to resolve steeper profiles. For future measurements, a reduced width of notch filter and intermediary slit in the spectrometer could yield additional spatial (1 mm) and spectral (0.2–0.3 nm) accuracy by using only the central row of fibers (cf. Fig. 6.4 right) and thus only a third of the photons. However, using the smaller notch filter allows receiving more of the central spectrum. The shape of the EEDF could then be analyzed in detail and deviations of a Maxwellian distribution would be detectable. Moreover, the interpretation of the electron velocity profile benefits from the increased spatial and spectral resolution.

The rough velocity profile estimates are comparable to previous findings at PSI-1 and PSI-2 in various gases, where mode structures with frequencies ranging from 10 to 100 kHz were found [13, 14]. Therein, the ion velocity profile was measured spectroscopically and reached about 2 km s^{-1}. The diamagnetic drift velocity $v^* = \frac{\nabla p}{e n B}$ based on the TS spectra result in similar figures, while for an overall rotation velocity the Langmuir probe derived plasma potential would be required. However, the uncertainty of the azimuthal plasma symmetry and the discrepancies between both diagnostics make this estimate unreliable and hence it is not presented. Nevertheless, investigating the plasma rotation is basically possible with the TS diagnostic, especially with higher resolution settings. Further aspects of the plasma dynamics are aided by the fast cameras in the next chapter, which help to put the high radial velocities and corresponding frequencies into perspective.

References

1. Waldmann O, Fussmann G, Bohmeyer W (2017) Ion mass spectrometry in a magnetized plasma
2. Reinhart M, Kreter A, Buzi L et al (2015) Influence of plasma impurities on the deuterium retention in tungsten exposed in the linear plasma generator psi-2. J Nucl Mater 463:1021–1024
3. Kreter A (2011) Reactor-relevant plasma-material interaction studies in linear plasma devices. Fusion Sci Technol 59:51–56; In: 8th international conference on open magnetic systems for plasma confinement (OS2010), Novosibirsk, Russia, 05–09 July 2010
4. Waldmann O, Meyer H, Fussmann G (2007) Anomalous diffusion in a linear plasma generator. Contrib Plasma Phys 47(10):691–702
5. Waldmann O, Fussmann G (2008) Influence of the langmuir probe shaft on measuring plasma parameters. Contrib Plasma Phys 48(5–7):534–539
6. Lunt T, Fussmann G, Waldmann O (2008) Experimental investigation of the plasma-wall transition. Phys Rev Lett 100(17):175004
7. Kreter A, Brandt C, Huber A et al (2015) Linear plasma device psi-2 for plasma-material interaction studies. Fusion Sci Technol 68(1):8–14
8. Reinhart M, Pospieszczyk A, Unterberg B et al (2013) Using the radiation of hydrogen atoms and molecules to determine electron density and temperature in the linear plasma device psi-2. Fusion Sci Technol 63:201–204
9. Sorokin I, Vizgalov I, Kurnaev V et al (2016) Studies of ion charge state distribution in noble gas plasmas on linear plasma device psi-2
10. Kastelewicz H, Fussmann G (2004) Plasma modelling for the psi linear plasma device. Contrib Plasma Phys 44(4):352–360
11. Klose S (2000) Untersuchung der Driftinstabilität an der rotierenden magnetisierten Plasmasäule des PSI-1 im Falle eines Plasmahohlprofils und großer endlicher Ionengyroradieneffekte. PhD thesis, HU Berlin
12. Langowski M (2009) Untersuchungen mit einer segmentierten neutralisatorplatte in einem linearen plasmagenerator. Master's thesis, HU Berlin AG Experimentelle Plasmaphysik

13. Klose S, Bohmeyer W, Laux M et al (2001) Investigation of ion drift waves in the psi-2 using langmuir-probes. Contrib Plasma Phys 41(5):467–472
14. Meyer H, Klose S, Pasch E, Fussmann G (2000) Plasma rotation in a plasma generator. Phys Rev E (Statistical Physics, Plasmas, Fluids, and Related Interdisciplinary Topics) 61(4):4347–4356

Chapter 8
Plasma Turbulence Results

Linear plasma generators are used to study basic plasma phenomena or with plasma as a tool to study plasma surface interactions in harsh environments, the harshest plasma environment by far being the first wall in a fusion reactor. Basic plasma dynamics entail studying plasma waves and instabilities and require the knowledge of many plasma and machine parameters to feed models and enable predictions. On the other hand, plasma as a tool should merely be of well known and adjustable density and temperature to serve as model input for understanding plasma surface interactions. But similar to the edge of a reactor, high plasma pressures in confined spaces require plasma gradients, which in turn create turbulence. Hence, until the control of turbulent transport is understood, controlled and suppressed, turbulence must be considered when using plasma as a (precision) tool.

By analyzing the plasma with the described diagnostics, the nature of the instabilities is sought after and compared to dynamics found in previous measurements at PSI-2 and other, similar plasma devices and also to theoretical descriptions in general. The plasma dynamics of the discharges is evaluated first with the fast framing cameras and spectral analysis. The conditional averaging method is applied connecting Thomson scattering and the fast camera to investigate transient plasma phenomena and intermittent filaments. Finally, the effects of temperature fluctuation changes induced by filaments on erosion are estimated with the statistical properties found by conditionally averaged measurements.

8.1 Plasma Dynamics

The dynamic evolution of a plasma in space and time is governed by micro-instabilities driven by the gradients of plasma density, temperature, flow and potential. Therefore, multipoint or profile measurements with sampling rates exceeding

M. Hubeny, *The Dynamics of Electrons in Linear Plasma Devices and Its Impact on Plasma Surface Interaction*, Springer Theses,
https://doi.org/10.1007/978-3-030-12536-3_8

the dynamic changes of the plasma parameters are required for classifying the instabilities. Two fast cameras provide versatile access to visible light emission profiles at different spatial positions. In conjunction with the Langmuir probe, the first aim is to analyze the dynamics in the PSI-2 plasma and investigate, whether the plasma edge shows turbulence features similar to the edge of other magnetized plasma devices. For a number of linear devices turbulence with plasma ejections have been reported [1–3] with time scales in the μs range.

A thin line of plasma with a few millimeters width is imaged (cf. Sect. 4.3) with the fast cameras, since the reading speed is optimized for one direction. The maximum frame rate of 1.4 MHz is achieved at the smallest resolution of 128×8 pixels, but the photon flux is rarely sufficient to create enough signal for a significant analysis. A compromise of lower frame rates and using 16 or 24 pixels generates usually enough signal intensity, with the standard deviation along these pixels as a measure of statistical error, since the brightness is assumed to be constant over a few millimeters in axial direction. Therefore, the collected frames are always axially averaged before processing and only brightness profile evolutions are presented in the following.

Examples of brightness profiles for discharges with all gases of the previous section are shown in Fig. 8.1 for a detailed time frame of 100 μs and a compressed frame with 300 μs. The right column depicts the average brightness profile above the (light) blue shaded area indicating a fluctuation level of ($2\times$) the standard deviation σ. The discharges gases are again ordered with decreasing atomic mass downwards. Fourier spectra will be shown (cf. Fig. 8.3) after a general description of the dynamics and after testing the validity of performing a Fourier analysis on projected profile evolutions.

The slowest brightness changes are found in the first two Argon discharges (Fig. 8.1a, b), where multiple, seemingly rotating structures are clearly visible at roughly the same frequency, called modes. The two dashed, sinusoidal lines trace particularly bright rotating modes, illustrating the onset and end of these modes. Besides the traced modes, there are a number of additional modes in the background, faintly visible if they appear isolated or as increased brightness when two or more modes cross in line-of-sight. As none of the other modes have a stable phase relation to the traced ones, these are considered $m = 1$ modes, where m denotes the azimuthal mode number. The turning points of the modes mark and constitute the outer limit of the brightness profile, while the lifetime of the modes is in the order of the oscillation period. Moreover, for both Argon discharges and depicted time scales the incoherence of the prevailing modes is apparent during the short times of strongly reduced mode activity. This complicates the identification of higher mode numbers, especially since the there is no fixed phase relation between multi-modes structures.

In the presented Neon discharge the high photon emissions allowed recording at almost the maximum frame rate of the fast camera, while the rotating modes have drastically increased in frequency and are thus only barely resolved. The structure radii have also widened from ± 3 cm to ± 5 cm. This results in a different radial range in Fig. 8.1c–e, focusing on the region with dynamic changes. Considering the

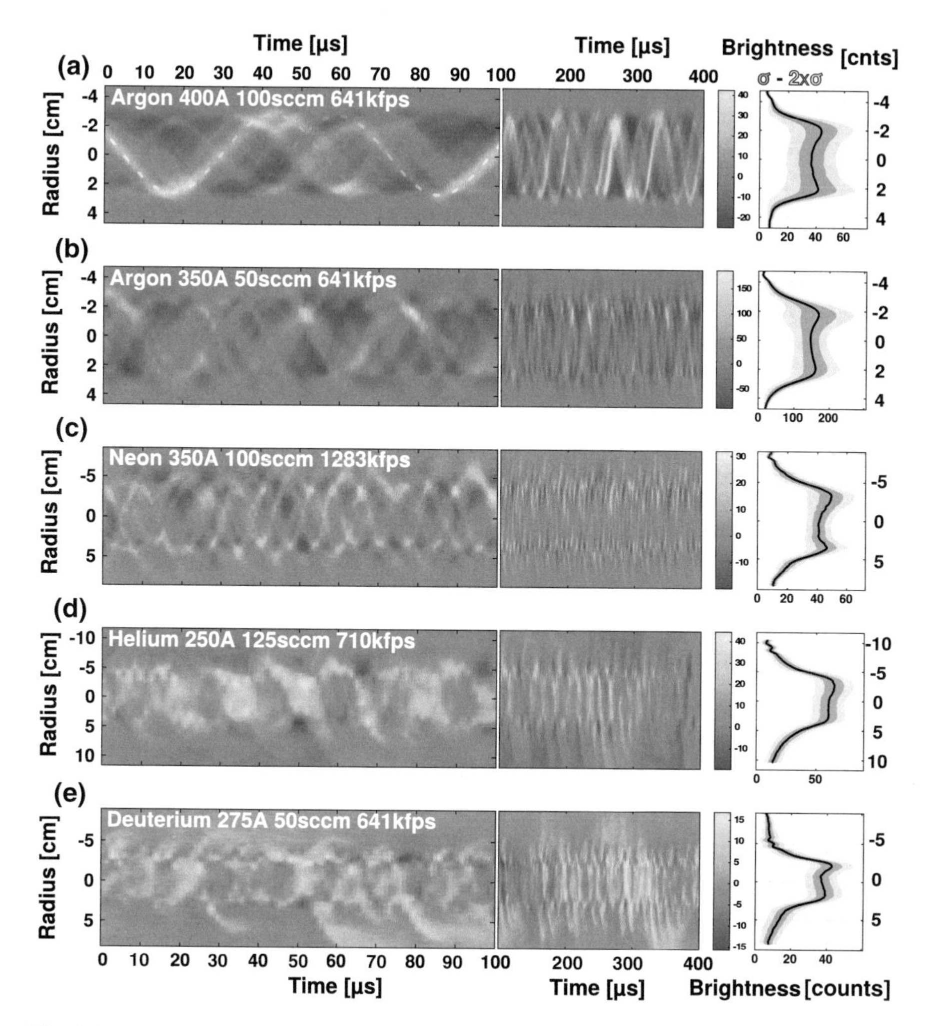

Fig. 8.1 Brightness profile evolutions for all tested gases on two time scales, showing faster dynamics with decreasing ion mass from **a** Argon to **e** Deuterium. The mean brightness is compared to the fluctuation strength on the right. The different radial extent is restricted by the frame size or the visible plasma column

number of crossing modes and occasional symmetric changes, higher mode numbers seem to be present.

The total brightness for Helium and Deuterium discharges is insufficient to allow recording at higher frame rates, which could resolve the fast dynamics in the bulk of the plasma in detail. Although the average profile and fast dynamic radii shrink slightly compared to the Neon case, larger structures are transported towards the far edge ($r \geq 8$ cm). While in Helium these structures are rather protruding up to a radius of ± 10 cm, in Deuterium they begin to disconnect from the bulk dynamics at a radius of about ±4 cm to ±5 cm. At the same radius, the mean brightness signal marks a slower decay, especially in Deuterium. In case this brightness corresponds to similar plasma parameters as the central region, the fast oscillation in the bulk plasma cause transport events reaching up to a radius of 10 cm.

Although integrating over a wide wavelength range (cf. Fig. 4.4) every gas type features distinct, rotating structures, with increasing velocity as the atomic mass of the ions decrease. The dynamics change also with variations of input power within the same gas type (cf. Fig. 8.22), while the here (Fig. 8.1) presented discharges were all high powered (20–30 kW). Similar to the density and temperature profiles, the brightness profiles are slightly hollow and some show an asymmetry in their maxima and decay length at the edge. Albeit an average hollow brightness profile could follow from the steady state hollow pressure profile, a rotating local emission source paired with a high level of fluctuation leads to another explanation: the projection of a rotating structure causes higher brightness at the turning point due to the slower projected angular velocity. Since the fluctuation amplitude and mean brightness are in the same order of magnitude for Argon, the mean profile is mostly produced by the rotating structures. Consequentially, they are visible at all times of their trajectory. The relative (to the mean) fluctuation amplitude decreases with the atomic mass, which could either be caused by smaller parameter fluctuations or a background consisting of unresolved fast fluctuations.

8.1.1 Fourier and Mode Analysis

Magnetized plasmas exhibit a plethora of wave phenomena with spatiotemporal domains spanning several magnitudes and thus there are virtually always oscillations present in a given plasma environment. Dominating growth rates within a given wave spectrum channel free energy to particularly excited wave modes, while still allowing complex fluctuation patterns, e.g. the predator-prey-model [4]. The adoption of this theoretical description to a plasma leads to at least two dominant plasma modes growing and decaying in amplitude, alternating between two unstable states. Different azimuthal mode structures are already visible in Fig. 8.1, while the structure and relation between the modes, life time and other statistical properties help to understand their generation. Since limited access to plasma parameters can obscure the recorded dynamics the mode structure is analyzed with regard to the limits of the diagnostics.

The verification, whether the projection of an oscillating mode structure introduces distortions, was performed for a number of structures of which two are shown as examples in Fig. 8.2. The background of the test profile consists of an angular symmetric and Gaussian profile peaking at 1.75 cm, thus creating a hollow profile, and the same profile with a sinusoidal angle dependence, which is rotated on a grid with 150×150 spatial points and 1 μs time resolution. The most important effects of the projection to consider are the $(1/r^2)$-power dependence of each light source, where r is the distance of each grid point source to the camera, and the circle-of-confusion,

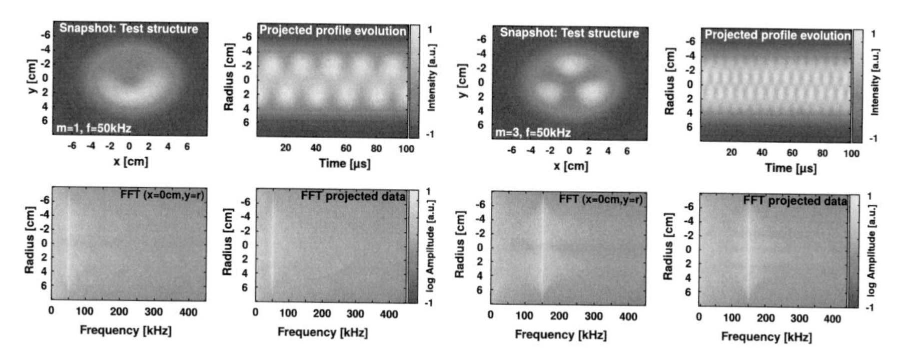

Fig. 8.2 Synthetic test of FFT for two rotating sample structures

which distributes light outside the focal plane (depth-of-field) over several pixels in the fast camera.

With these two effects the rotating $m = 1$ maximum in the projection is sharply imaged and thus appears brighter at the edges, while connections in-between are slightly smeared out for being out of focus. The Fourier transformation in the focal plane is compared to the projected profile evolution and shows no distortions. The right example in Fig. 8.2 shows an $m = 3$ mode structure at the same rotation frequency, with the brightness maxima being at the times when two maxima overlap. Here, only the threefold frequency appears in the Fourier analysis of the projection. Further tests with slow and fast structures, frequency sweeps (broader frequency peak) and randomly seeded wave trains were conducted to test the response of the FFT to these projections. Modes with different structures and frequencies can be separated and are well recognizable in the frequency space. Therefore, the main features of oscillations are assumed to stay intact and can be distinguished by fast camera measurements from a side projection.

Simulating the partial re-absorption of emissions along longer optical paths was abandoned since the effect is assumed to be small but needs proper three dimensional modeling of the light path through the plasma structures towards the collection optics.

Real examples of normalized power spectra are now compared in Fig. 8.3 for Argon, Neon and Deuterium adjusted in radius and with individual frequency scales each to accommodate the important features. The velocity increase with ion mass is partially visible only, since the overlap of multiple slow modes in Argon causes the appearance of harmonics, while the fast oscillations in Neon seem to create several distinct modes at 43, 84, 128, 166 kHz and an additional slow mode at $f = 9.4$ kHz. In Argon and particularly in Neon the spectral power maxima are radially localized with nodes of lower power moving outwards (indicated by dashed line), but the overall extent of the maxima are located around the pressure maxima. The power spectrum for Deuterium shows only one broad peak around 80 kHz and the slow, far reaching oscillation at 7.5 kHz, which is also seen in the conditional averaging (cf. Fig. 8.9). The faster peak is again split by a node at $r = \pm 2.5$ cm, but a major difference in the power spectrum is a high spectral power mode at low frequencies and the long

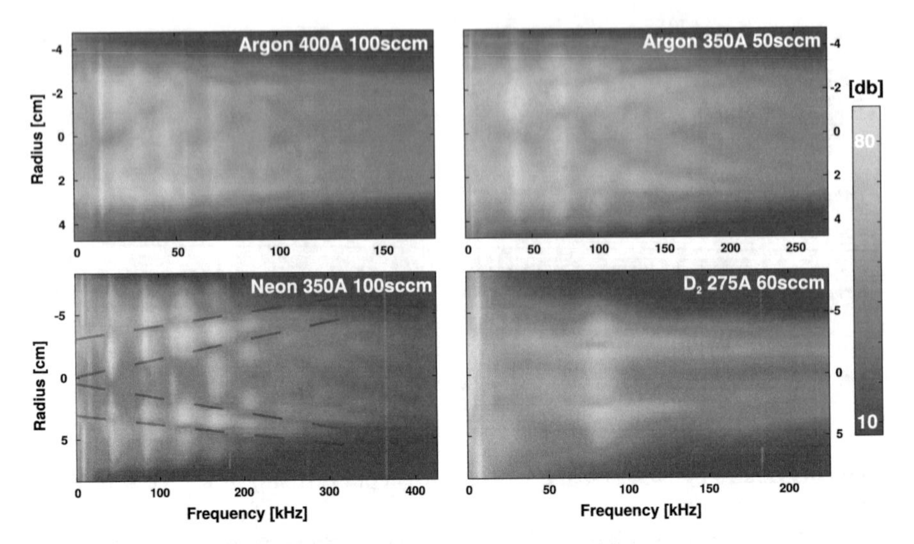

Fig. 8.3 Radially resolved power spectrum (logarithmic intensity) of four discharges. The frequency range is chosen to accommodate the mode structure frequencies

tail towards higher frequency. This could either be caused by broadband turbulence or faster, unresolved dynamics. The photon emissions in Deuterium discharges are insufficient for faster exposures, but higher harmonics can be ruled out.

The standard diagnostic for fluctuation measurements are Langmuir probes and thus always used as an reference for other time-resolved diagnostic. Usually, the Langmuir probe signals at PSI-2 is digitized at 100 kHz for 10 s to cover the full profile traveling in and out. Subtracting a sliding average and extracting moving windows from the complete data set of the ion saturation current I_{sat}, a radial power spectrum can be computed, which is shown on the left in Fig. 8.4 for a Deuterium discharge. Although at lower discharge power, typical features are recognizable, namely the slow oscillation ($f = 7.5$ kHz) localized at $r = 6.5$ cm and increased spectral power close at the pressure maximum and zero frequency. The frequency response decrease beyond 30 kHz is likely caused by the bandwidth limit, while a decline of spectral power similar to the node in the fast camera power spectra is still visible at a radius of 2.5 cm.

Using the transient recorder HIOKI 8861 the Langmuir signals could be digitized at 1 MHz, but for only fixed length of 2 s and hence for a limited radial extent. The power spectrum for a sweep through the maximum pressure region of a Neon discharge is shown on the right in Fig. 8.4, compared with the radially averaged power spectrum from Fig. 8.3. Most of the features are well comparable, while the peak at $f = 9.4$ kHz is absent but appears in measurements at the plasma edge (not shown).

The comparison of the spectral power profiles obtained by Langmuir probe and fast camera measurements confirms the equivalence of both diagnostics for characterizing plasma fluctuations and observing fast transients, since the frequency response is

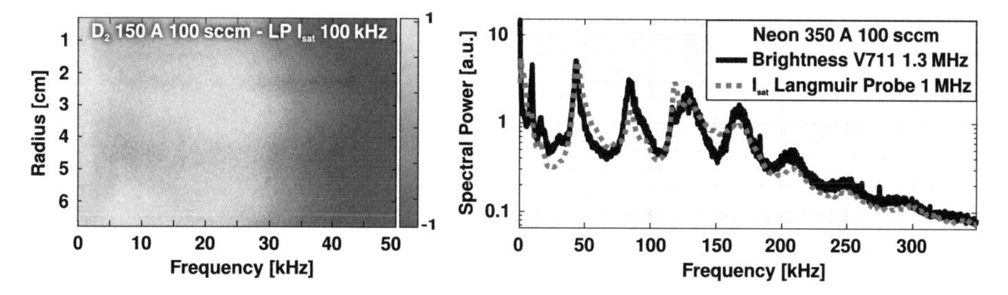

Fig. 8.4 Power spectrum of I_{sat} (moving window FFT) in Deuterium (left) and a radially averaged comparison between I_{sat} and brightness power spectrum for a Neon discharge (right)

identical towards higher frequencies with the transient recorder. While the Langmuir probe measures local profiles, the fast camera captures the whole plasma profile and benefits especially from optically thin plasma conditions, where fluctuations can be traced throughout a azimuthal plasma cross-section.

The axial length (parallel to magnetic field) of the modes is estimated by comparing simultaneous fast camera measurements at ports F2.1/F3.1 or F2.1/F4.1 with an axial distance of 40 and 80 cm, respectively. Since the plasma transport along the magnetic field is much faster than perpendicular to it, coherent structures are expected to stretch over this distance and result in high coherence ($\geq\sim 0.5$). Deuterium discharges ranging from 100 to 300 A and 50 to 150 sccm were imaged with 182k fps (v641) and 912k fps (v711). Discharges with lower photon emission were additionally recorded at 136k fps (v641) and 680k fps (v711). Depending on the reference signal, either the faster signal is down-sampled or the slower one interpolated.

Ideally, every fifth frame is recorded at an identical time for an exactly matching frame rate ratio of 5, but hardware restrictions limit the frame rate settings and hence the slow video frames are not exactly synchronized to every fifth fast frame. Therefore, the time relations of the temporally closest frames drift and a higher ratio is still beneficial for matching the slower time base for the necessary down-sampling and interpolation of the faster v711 video.

Figure 8.5 shows the cross-power spectrum (top) and coherence (bottom) for a 300 A, 75 sccm Deuterium discharge according to Eqs. 6.6 and 6.7 for about 110 and 6400 data points in radial and temporal dimension, respectively. The frequency resolution of the FFT is 178 Hz and overlapping windows with 1024 data points were used for a realization, but other settings were tested as well with similar results. Spectral power and coherence show high values for the low frequency (7.5 kHz), high frequency (80 kHz) and towards zero frequency in the center. The radial extent is restricted due to high sampling frequency, but all important spectral features are visible and lead to the conclusion of axially elongated structures. These results are comparable for all Deuterium discharges, while the spectral positions are slightly different, yet exhibit high coherence values. Furthermore, similar results are obtained for Neon discharges ranging from 100 A to 300 A and 50 to 150 sccm.

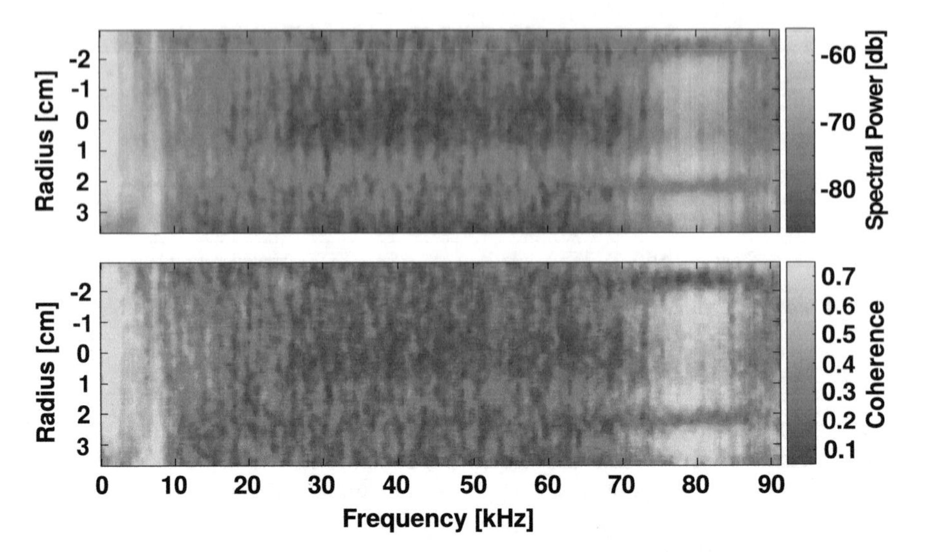

Fig. 8.5 Spectrally resolved cross-power P_{xy} (top) and coherence C_{xy} for a 300 A, 75 sccm Deuterium discharge with a distance of 80 cm (F2.1 and F4.1)

Mode Analysis

For two dimensional signal analysis, there are two principle ways of mode analysis: spatial or temporal. The amplitude evolution in the temporal sense can be realized by a moving FFT window much smaller than the total length of the signal, and averaging over the whole or part of the radius. If an oscillation is not well characterized by a single frequency, wavelet analysis uses predefined wave trains instead of sinusoidal functions. Although the modes in PSI-2 are sinusoidal, they last only a few cycles due to the short correlation time and overlap in time, e.g. in Argon and Neon discharges. This can be seen in Fig. 8.6, where examples of auto-correlations for Argon and Deuterium are shown at selected radii. Therefore, this representation is only a simplification from a full two-dimensional analysis of the decay lengths. However, the presented positions are chosen as representative samples in the mode structures (cf. Figs. 8.1 and 8.3) to avoid multiple, overlaying oscillations and are hence sufficient to show the general lifetime of plasma structures in the discharge conditions.

There are usually a slow and a fast or not resolvable component, recognizable by the immediate drop around zero time delay. While the central drop can also be explained by random fluctuations or noise, the large fluctuation amplitudes (cf. Fig. 8.1) point towards irregular occurring brightness excursions. The correlation time estimate is based on fitting an exponential function $A(t) = A \exp(-t/\tau_{1/e})$ with amplitude A and decay time $\tau_{1/e}$, where the amplitude is only set to accommodate the slow decay, which is in the order of 500 μs. The only resolvable fast decay is displayed in the upper right picture with a much faster decay of 32 μs, while other examples exhibit a rapid decrease of the auto-correlation coefficient within the first

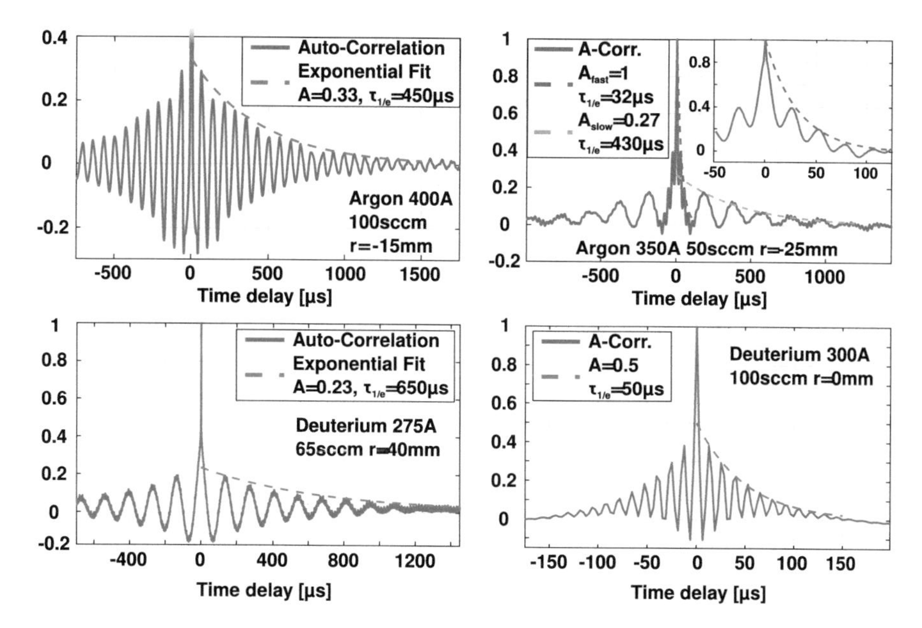

Fig. 8.6 Normalized auto-correlation examples for Argon and Deuterium at selected positions with exponential functions fitted to the envelope of the oscillation, disregarding the initial peak at $t = 0\,\mu s$

few μs, which points either to much faster time scales or singular events with high brightness.

Circumventing short correlation times, spatial mode analysis can be used for poloidal cross-sections, where the modes are retrieved by a spatial FFT along the radius. However, when imaging the oscillation and using only a projection for the analysis, the so-called principal component analysis finds the basic modes by eigenvalue decomposition of a correlation matrix. The largest eigenvalues denote the modes with the highest amplitude and the time evolution of these principal modes show possible mode interactions or dependencies.

Both spectral mode analysis with moving windows and spatial mode analysis with principal component analysis do not lead to a clear dependence between the oscillation amplitudes of various modes in the analyzed discharges. Short correlation times and the excitation of various modes and mode numbers point towards a broadband turbulent plasma state in which the observed modes are excited at random. The oscillation frequency and mode structures are in the range of drift-wave instabilities, which are created by the plasma pressure gradient and have been identified before at PSI-1 [5].

Drift-wave turbulence is one of the possible causes for filament observations in magnetically confined plasmas, while the generation and propulsion depends on the particular plasma environment. Both drift-wave and filaments are structures elongated along the magnetic field, which can be confirmed by the coherence at high spectral powers. The outbursts found in the edge of Deuterium discharges are thus identified as filaments, which, if they occur intermittently, could be responsible for the fast decay of the auto-correlation function in Deuterium. Therefore, the intermittent character of the fluctuations is usually investigated by statistical measures.

8.1.2 *Statistical Analysis and Conditional Average*

From a statistical point of view the momenta of the time series of brightness from the fast camera and the ion saturation current of the Langmuir probe often serve as the first indicator of intermittence. In Fig. 8.7 the mean (gray), standard deviation σ (black), skewness (red) and kurtosis (blue) are shown for the fast camera data of a high power Deuterium discharge (Fig. 8.1e). While the mean brightness is scaled, σ, skewness and kurtosis are shown unscaled. The minimum of the skewness is reached slightly outside the maximum brightness and increases inside and outside up to a value of $\text{S} = 2$, while the kurtosis shows a similar behavior. All curves show slight asymmetries, while the general features are similar on both sides. An indication of the hollow profile shape is visible in mean and σ of the brightness. The comparison between mean and σ shows a higher level of relative fluctuation amplitude towards the edge as the mean brightness decreases faster at smaller radii.

However, a direct normalization of the fluctuation amplitude ($\sim \sigma$) to the mean brightness is ambiguous, since the minimum brightness for high frame rates is elevated and subtracted based on the smallest average brightness in a movie. On the other hand, the skewness and kurtosis are statistical momenta normalized by the standard deviation and thus independent measures. The three positions marked with black dashed lines are marked for the extrema of the skewness profile and the center position.

The most important feature is the increase of skewness and kurtosis towards the edge of the plasma, while the maxima of the brightness profile are close to the minima of the skewness at about $r = \pm 3$ cm. The skewness and kurtosis of a Gaussian amplitude distribution is zero, while positive deviations show an over-abundance of

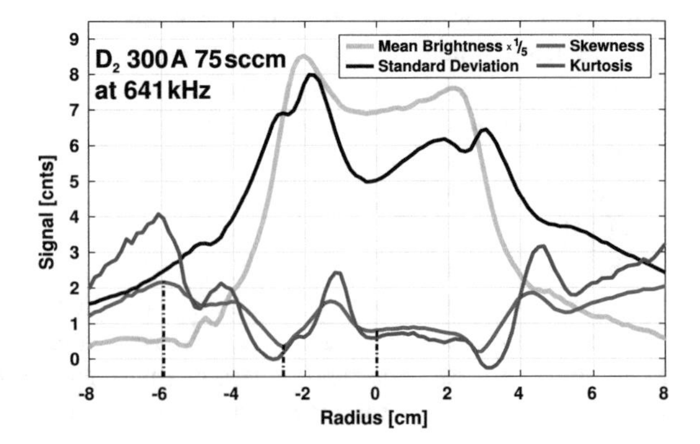

Fig. 8.7 Radial dependence of statistical moments of the brightness for a Deuterium (black) and Argon (grey) discharge. The red lines mark positions for Probability density function analysis of the Deuterium signal

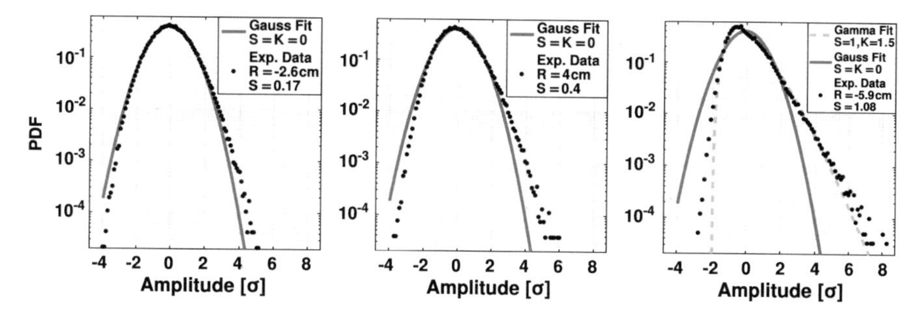

Fig. 8.8 Probability density function for a Deuterium discharge at three radial positions, compared to Gaussian and Γ distribution

large positive events (skewness) with a steep amplitude rise (kurtosis). This motivates another visualization of the non-Gaussian character of the fluctuation, the probability density function (PDF) of the brightness signal, shown in Fig. 8.8. Gaussian PDFs are compared to the actual distribution of amplitudes at three different radial positions (marked red Fig. 8.7). At the maximum brightness the skewness is low and hence the PDF deviates only slightly from a Gaussian distribution, while at the edge the PDF is heavily skewed with a long tail reaching several σ deviation.

In the time series of the brightness fluctuations normalized by σ the largest positive peaks are found in the edge. Without a clear resemblance in the frequency space, i.e. there is no typical average waiting time in-between the peaks, these events are characterized as intermittent. To further characterize their general properties the technique called conditional averaging (Sect. 6.6) is applied to the time series. Usually Langmuir probe data is analyzed, but since the spectral properties are comparable and the fast camera frames give access the entire brightness profile evolution, only these are are analyzed and presented in the following. Constructing the conditional average of the profile evolution requires a minimum of one brightness threshold at one radial position. Including a number of frames before and after the triggered event shows the state of the plasma (inferred from the brightness) around reaching the averaging condition. Four examples are shown in Fig. 8.9, where the trigger position is marked with a white "x" and the trigger threshold stated in each picture, as well as the number of events used for the averaging. The radial size of the trigger time trace is 3 mm and the minimum waiting time is set to 32 frames ($\approx$25 μs). This rejects fast successive events and sets the minimal waiting time in-between events. The single-sided conditions (a, b, d) show a common brightness evolution in which a slow and far outreaching oscillation is always roughly in phase with the detected event, while a rapid oscillation in the bulk plasma is growing in intensity about 50–100 μs prior to the event.

The dashed white line follows the slow oscillation and reveals a slightly different timing between trigger and slow oscillation phase in these three cases. While the upper events sits right in the maximum, the lower trigger point is ahead of the slow

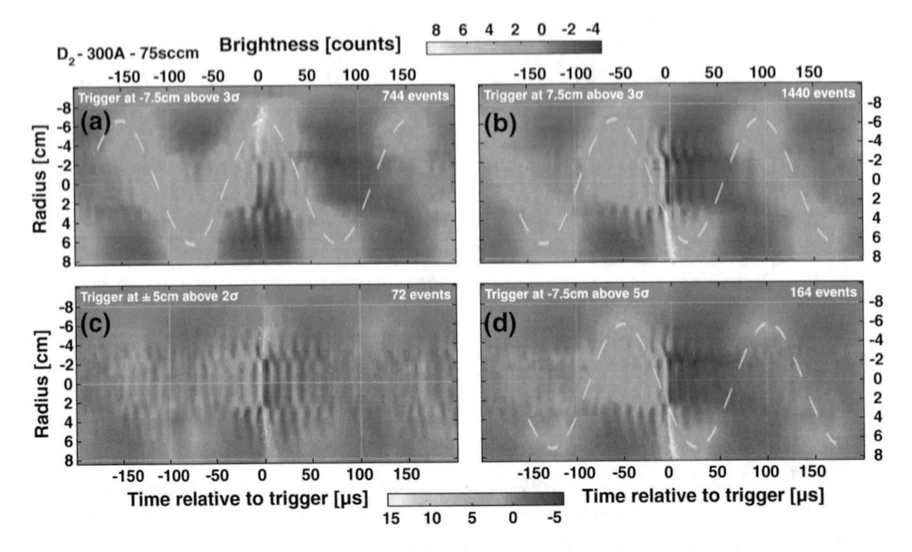

Fig. 8.9 Conditional average for upper, lower and both-sided events with marked trigger radii. The color bars correspond to both pictures in each row

oscillation maximum. Furthermore the lead seem to increase with higher trigger threshold, suggesting a faster growth rate of the bulk instability.

The ratio between upper and lower side event numbers is about $744/1440 \approx 2$ as seen in the figure for 3σ, 238/540 for 4σ and 110/164 for 5σ. Although the slow oscillation seems unaffected by the events, the lower outbursts cause a distinct drop in bulk brightness for about 50 μs indicating a larger impact on the plasma column. The brightness at negative radii declines abruptly at about $r = 8.5$ cm even at higher trigger thresholds, while triggering at larger negative radii only yields matches at low thresholds and shows increased bulk brightness likely caused by reflections. Thus, the total extent of the outburst on both side is limited by the optical imaging (view field) of the camera. While some events show slowing or change of trajectory backwards, the larger events resembled in Fig. 8.9d are faster, more compact and are likely to reach a projected 10–12 cm radius.

Setting two triggers at opposite radial positions is tested in Fig. 8.9c, where simultaneous outbursts are averaged. Here, the slow rotation is absent, but the fast oscillation is excited similar to the previous cases. The probability of these events is about a factor 10–20 lower, even at a threshold of 2 σ. This low event number prohibits a detailed investigation in this work, since the number of photons captured by the TS diagnostic roughly correlates to the number of events, and at least 2000–2500 events are needed for a analysis (cf. Fig. 8.23). However, the occurrence and clear differentiation between single and double-sided events shows that the slow rotation is not a necessary condition for an outburst.

The slope of the rising brightness front corresponds to either a radial velocity or a projected and thus higher, azimuthal velocity in the range of several km/s. To

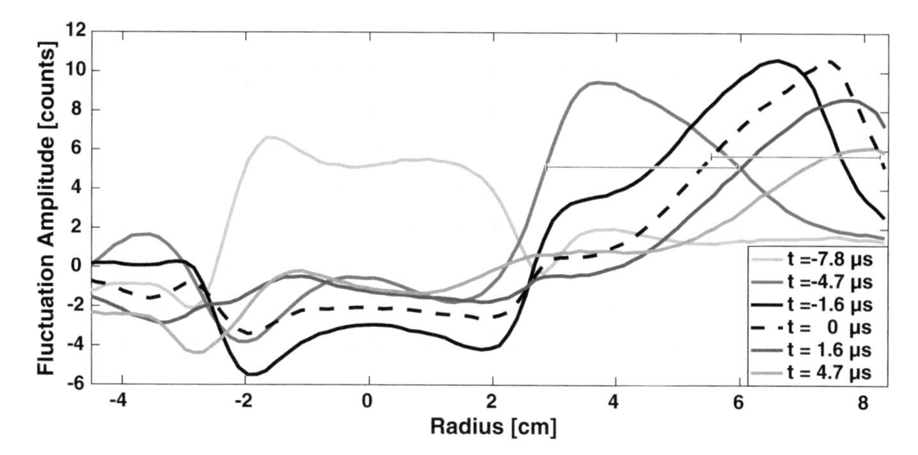

Fig. 8.10 Profile evolution during an outburst, indicating spatial size and projected velocity of the coherent structure around the trigger time

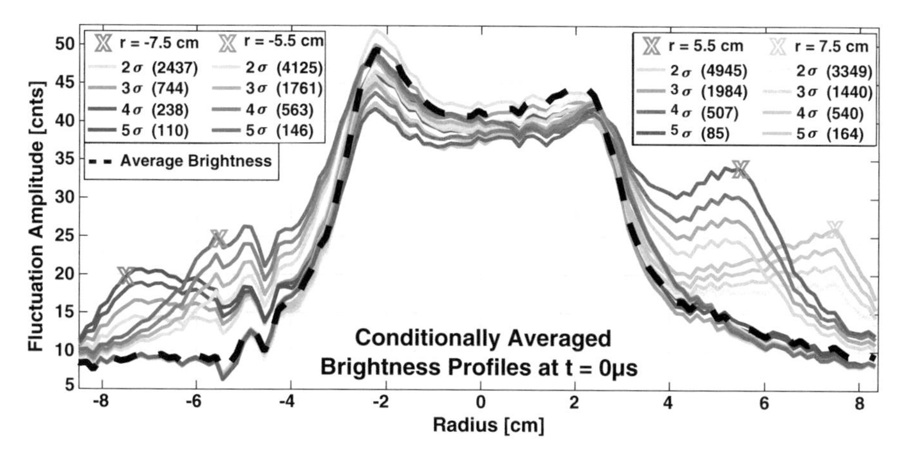

Fig. 8.11 Absolute intensity profiles of various trigger positions, indicated by color-coded "x", with four thresholds each. The number of captured events is noted in brackets and the average brightness profile shown as dashed black line

estimate some key properties for this specific set of conditions, two figures are shown to illustrate the calculation methods in order to depict these properties depending on the plasma discharge settings. Selected radial brightness profiles without background are shown in Fig. 8.10, where tracking the brightness onset translates to a radial velocity of 5–7 km/s (7–15 km/s azimuthal) and the FWHM (dotted line) in spatial and temporal dimension is $\Delta r \approx 3$ cm and $\Delta t \approx 3-4\,\mu$s, respectively. The actual size is about 1 cm smaller considering the movement of at least 7.5 mm (at 5 km/s) of the structure during the frame duration of 1.6 µs. Note the different time intervals, which were chosen to highlight the essential profile changes.

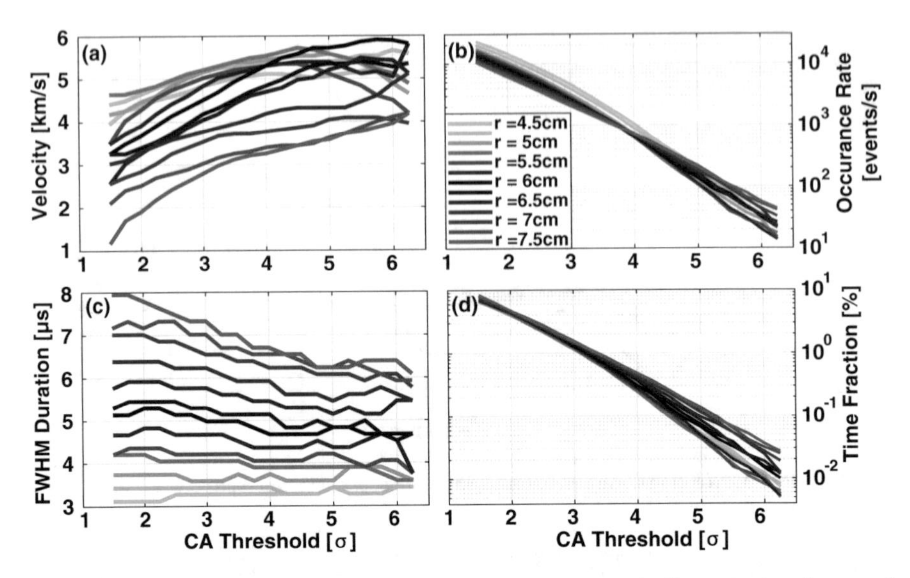

Fig. 8.12 Results of CA threshold and CA trigger radius scan: **a** velocity at ejection, **b** counted events per second, **c** FWHM time duration for each event and **d** probability of finding event in frame

The conditionally averaged brightness amplitudes at trigger time compared to the mean background profile are shown in Fig. 8.11, where three trigger positions with four trigger thresholds each are clearly visible as detached structures at $t = 0\,\mu s$. Furthermore, the occurrence at each threshold is indicated in the legend. Similar to Fig. 8.9 the events at negative radii are occurring slightly less frequent and on both sides the occurrences decrease towards larger radii. From the profile evolution in Fig. 8.10 it is clear that triggering at different radii will often find the same events, just at different times on their trajectory, which must be considered when estimating the likelihood of occurrence.

To compare the properties of averaged events to theoretical models the dynamics and intensity scalings are essential. Based on the general observations of Figs. 8.9 and 8.10 the directly accessible properties are now shown comprehensively in Fig. 8.12 for variations of trigger thresholds and radii. The velocity is estimated in Fig. 8.12 a by averaging the delay between the two adjacent frames upon ejection and additionally the delay of their cross-correlations to account for changing shapes, making the estimate more robust. Generally the velocity scales positively with intensity (σ) and events detected at larger radii have a lower initial velocity, but increase in velocity more rapidly while high intensity events near the bulk plasma reach a saturation velocity between 5 and 6 km/s. Since the velocity is estimated around the trigger radius these dependencies are valid only for the particular position and fit the picture of a rapid ejection close to the plasma edge and then the dependence on the intensity increases as only high intensity events reach the far edge. The apparent slowing of smaller events could be caused by friction or a change in trajectory.

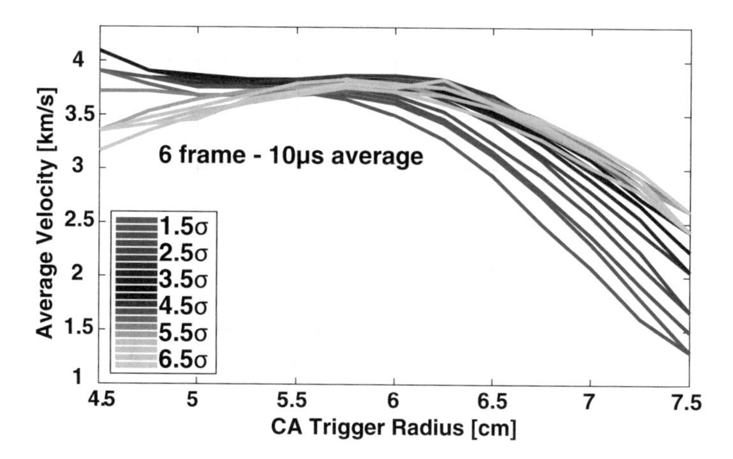

Fig. 8.13 Average velocity vs trigger radius over 6 frames (≈10 μs)

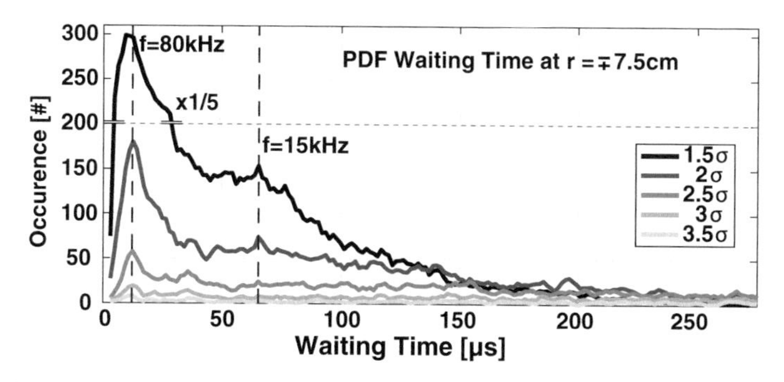

Fig. 8.14 Waiting time at $r = \pm 7.5$ cm combined

The rate at which events are counted is shown in Fig. 8.12b and decreases sharply from several 10^4 to a few tens of events. Smaller events are more likely to occur closer to the bulk while larger events are more likely in the far edge, i.e. these large events make up more of the total signal with increasing radius. The duration dependence on radius and intensity is shown in Fig. 8.12c, where the duration is shortest and roughly independent of σ for small radii and increasing with larger radii. As the trigger radius increases the duration begins to shorten with increasing intensity.

In Fig. 8.12d, the combination of duration and occurrence rate shows the total amount of time at which the plasma is in the state of an ejection. Here the dead time between counting events must be reduced to a few μs and the counting routine includes ejections in rapid succession if the amplitude drops to half its height in-between.

Another representation of the ejection velocity is shown in Fig. 8.13. The 10 μs velocity average is roughly constant for all trigger thresholds, while towards the edge events with higher intensity stay faster and the velocity small events declines.

The waiting time between events coined the term "intermittent" for the description of the plasma ejections in the edge of plasma devices for there are no typical average time delays between following events. In Fig. 8.14 the far edge of the PSI-2 plasma shows an almost monotonically declining waiting time for small events and a flat distribution for larger events. However, there is one large peak at 12.5 μs and a small peak around 65.5 μs, corresponding to a frequency of 80 kHz and 15 kHz, respectively. The rapid oscillation located within the bulk plasma is setting the minimum time between ejections, causing this waiting time to be most likely, yet the peak is rather broad and the ejections are neither totally random nor are they appearing at a steady rate. Therefore, the conditional averaging method gives likely a better insight to the dynamics, since it automatically aligns the fluctuations leading to these events.

The analysis of structure and dynamics of the events, found by setting a conditional trigger with variable threshold, leads to the conclusion that there are coherent structures with a large velocity propelled out from the bulk plasma. The size of 2–3 cm and a velocity of several km/s fits well to filament properties found in the literature [6]. The conditional averaging connects the filament observation in Deuterium discharges to a bulk oscillation, while a slow outer oscillation seems to increase filament ejection probability, but is not a necessary condition.

The brightness evolution recorded by the fast camera(s) is a versatile tool to get an impression of the plasma dynamics and is applicable to all tested plasma conditions. Statistical properties can be obtained for a full plasma profile and the results can be considered comparable to those obtained by the Langmuir probe. However, brightness and ion saturation current I_{sat} suffer from the ambiguity of its signal in terms of relating them to plasma density and temperature fluctuations. While the relation for $I_{sat} \sim n_e\sqrt{T_e}$ is a good approximation, the general photon emission from a plasma originates from various processes, e.g. line radiation or collision induced. The emission depends on the photon energy, neutral pressure, plasma density and temperature. Although for a limited plasma parameter range and a dominant process, e.g. line radiation in Deuterium, this relationship can reduce to just brightness $\sim n_e$, generally not even a linear dependence can be assumed. Lastly, the fast camera integrates the photon flux over a wide spectral range, which is uncritical for Deuterium and its dominant H_α emission, while Argon emits at a number of wavelengths with each different parameter dependencies.

These experimental findings and interpretation limitations strongly motivates the verification of the found dynamics by a diagnostic method, which differentiates density and temperature, but is also time resolved. The connection between TS and the fast cameras with the conditional averaging provides this possibility, and the spatial resolution of TS setup is high enough to resolve the coherent structures in Argon and Deuterium.

8.2 Conditionally Averaged Thomson Scattering

The final goal of achieving time resolved measurements with laser scattering on low density plasma is concluded with the combination of fast camera measurements accurately synchronized with the Nd:YAG laser and the TS spectrometer. The test case with slow oscillations and high brightness amplitudes is the Argon discharge with 350 A and 50 sccm, since the laser beam and stray-light conditions are stable and there is a dominant oscillation, which can be investigated to test the diagnostic system. Although there are no intermittent plasma ejections in Argon discharges, the short life time of the oscillations and thus the limited coherence time are an excellent test case, since the changes occur in the plasma peak where the sensitivity is high and the oscillations are always present and just have to be "sorted" by their phase relative to the trigger and laser pulse.

From the view point on the plasma dynamics, the Argon discharges are the opposite end with the highest ion mass of all the used gas species, highest densities profile and the lowest observed frequencies.

The analysis of the conditionally averaged Thomson spectra is inherently more inaccurate (statistically) since only a fraction of the total signal length is used and thus less photons are available for a TS spectrum for the same discharge time. Instead of spatial smoothing the resulting parameters, a different interpolation approach is increasing the spatial binning to 2 cm, but shifting the spatial channel by only 0.5 cm (overlapping 1.5 cm with each shift). Although the spatial resolution is reduced to 2 cm, existing smaller structures or gradients are only dampened, while the Gaussian fitting routine works more robust without failing the fitting process due to sparse photon numbers.

Figure 8.15 shows a comparison between an equal native and effective resolution of 0.5 cm, while the black curves use 2 cm of data. The error bars here present only the fitting error without the statistical errors since the latter remains unchanged. This emphasizes the effect of the increased fitting accuracy, based on the residual of the Gaussian fit. The smoothing effect is clearly visible in maxima and slope of density and temperature profile, while the changes remain mostly within the fitting error bars. The modes in Argon are in the order of 1–2 cm, while the filaments in Deuterium are 2–3 cm, thus the resolution after the reduction is still high enough to capture the structure in just one spatial channel.

8.2.1 Coherent Oscillations in Argon Discharges

For the conditional average analysis in Argon 14,000 out of 21,000 laser shots were recorded while synchronized with the monochrome fast camera without spectral filters. The steady state discharge profiles were presented in Fig. 7.4 and the dynamics analyzed by fast camera data in the previous section. The cross-referencing of Thomson scattering and brightness profile evolution is now tested.

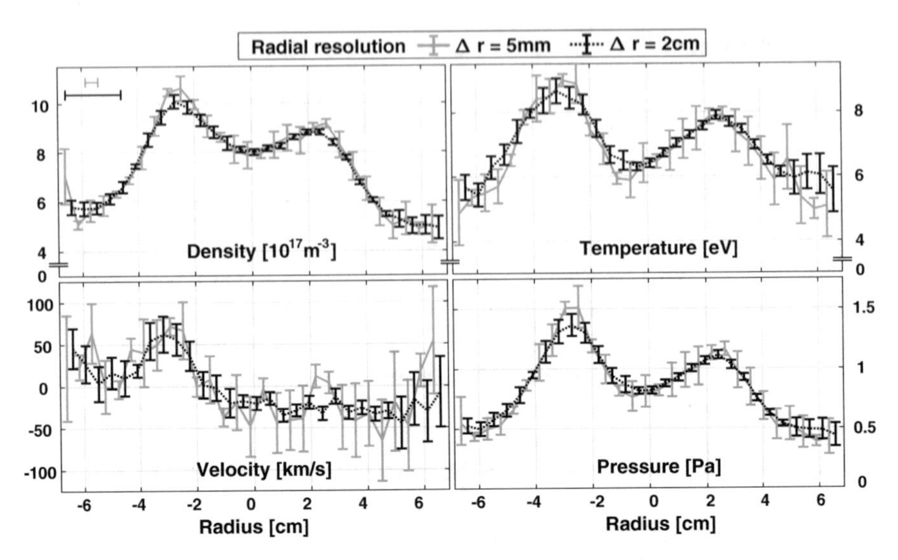

Fig. 8.15 Thomson scattering profiles compared for two spatial resolutions with error bars representing only the fitting error, without the statistical error

While the iCCD recorded frames with and without laser pulse at combined rate of 20 Hz, the fast camera recorded 100 frames at $f_s = 333$ kHz around each iCCD trigger, where the 55th frame coincides with (and records) the laser pulse (in every second set). This is somewhat slower compared to the analysis in the previous section but a necessary compromise between the amount of data, data analysis and the time resolution. For the slower oscillations in Argon and specifically the prominent, identified m = 1 mode at the 350 A/50 sccm settings, the number of frames and the resolution is sufficient to identify the phase of the oscillation with respect to the trigger frame. Since there are 14,000 shots, with each 100 frames and 128 pixels (radial), the trigger condition can be varied in time and space. To follow the onset and decay of the oscillation the conditional trigger threshold was kept at a constant $\sigma \leq 1.25 \ldots 2$ and radius $r \approx 2$ cm, while the trigger frame was cycled from $t = -120 \,..\, 120\,\mu$s with a time step of $\Delta t = 1/f_s = 3\,\mu$s.

To illustrate the generation of the subsequent two-dimensional profile evolutions, five brightness evolutions of conditionally averaged ($\sigma \leq 1.25$, $r \approx 2$ cm) fast camera frames are shown on the left in Fig. 8.16. As the trigger frame is increased, the maximum of the oscillation shifts relative to the laser-containing frame. Setting the conditional trigger in the laser frame leads to an increase in brightness, which is not averaged out, while other trigger times show almost no sign of the laser signal. This is likely caused by the overall increased brightness due to the laser and slight offsets in subtracting the mean.

The Thomson profiles of density, temperature and pressure are shown on the right for the central three cases. The time resolution of Thomson scattering is thus created relative to the conditional trigger time, which is indicated on the left with a

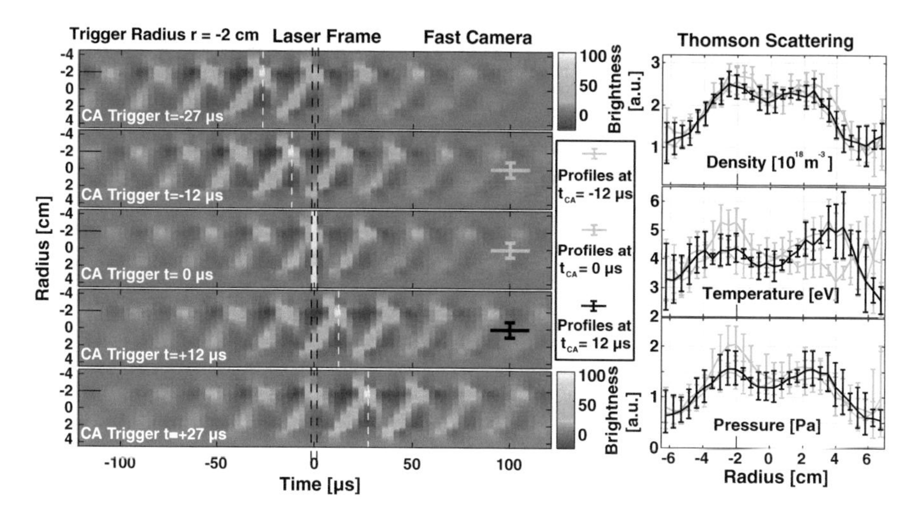

Fig. 8.16 Generation of Thomson scattering profiles triggered by CA brightness profiles in Argon. The left side shows five brightness profile evolutions trigger at different times (indicated) and at $r = 2\,\text{cm}$. Thomson profiles, obtained at the laser mark, are shown on the right for the central three brightness evolutions

dashed line. The trigger times in this example are chosen to match the minimum and maximum of the brightness oscillation. The resulting Thomson scattering density profile for $t = 0\,\mu\text{s}$ shows a slight increase compared to the other two cases in the whole plasma but mostly within the error bars. In contrast, the temperature profiles follow the brightness clearly and the changes are significant relative to the error bars. The pressure profile reveals, that the oscillation only increases the pressure significantly at positive radii, while the pressure change at negative radii is small due to the increased density counteracting the temperature drop.

To obtain the complete density and temperature profile evolution all trigger times are combined and compared to the average brightness evolutions, which are added with respect to its trigger time to average out the increased brightness during the laser frame. Figure 8.17 presents a comparison of the brightness evolution for the condition of a $\sigma \geq 1.25$ at $r = 1.6$–2.2cm and the corresponding density and temperature profile evolutions. The resemblance of brightness and temperature fluctuation is striking, while the density evolution shows some degree of oscillation around $t = \pm\, 50\,\mu\text{s}$ with its highest increase at the same radial position compared to the brightness. The maximum of the temperature oscillation amplitude is shifted slightly outwards to $r \approx 3\,\text{cm}$. The oscillation in T_e and brightness is matching in phase and decay of its amplitude and while imaging the brightness reveals the complete, seeming rotation, visible in the center of Fig. 8.17, the T_e profile lacks the connection between the maxima. This is caused by the 6 cm depth-of-field of the fast camera, which images the rotation also out of the laser plane, while the TS signal is exclusively emitted and collected from the laser beam diameter.

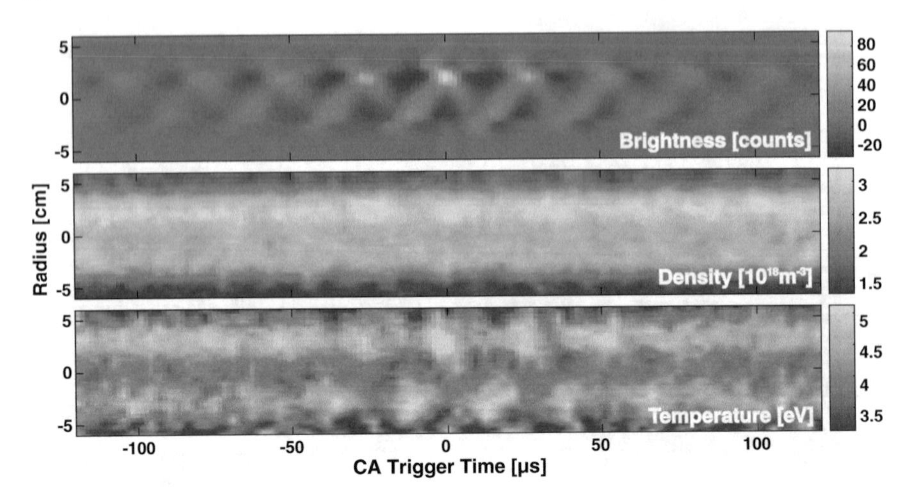

Fig. 8.17 Brightness, density and temperature profile evolution for a conditional average with $\sigma \geq 1.25$ at $r = 1.6$–$2.2\,\text{cm}$ for an Argon discharge

To highlight the correlation between brightness and temperature/density, Fig. 8.18 shows traces of the parameters at their upper and lower peak amplitude radii, respectively. In the upper picture, the brightness is normalized by σ and reaches a slightly less value then the threshold, since the shown trace is covering a range of one cm. The amplitude of the temperature oscillation is about 1 eV corresponding to 20 % of the $T_e = 4.5\,\text{eV}$ average value at the upper profile maximum. The period between the maxima is approximately $25\,\mu\text{s}$ and thus consistent with the base frequency of $\text{f} = 40\,\text{kHz}$ shown in Figs. 8.3 and 8.1. The correlation time of the oscillation is estimated to $t_\tau = 75\,\mu\text{s}$ by the roughly fitted Gaussian envelope function.

Although the density profile misses a clear phase resemblance with the brightness evolution, a significant increase by about $2 \times 10^{17}\,\text{m}^{-3}$ is found in the upper half of the profile ($r \approx -2.3\,\text{cm}$). Two density excursions at $t = 0$ and $-20\,\mu\text{s}$ seem to be in phase with the temperature and brightness, while following peaks are then unrelated to the oscillation. An unstable phase and/or frequency relation between brightness/T_e and n_e oscillations would average out the density fluctuations in the process of CA subset creation. Only the initial peak(s) would be retained and may indicate an unsuccessful coupling between modes.

The temperature fluctuation found in the Argon discharge served as a test bed for applying the conditional averaging method the Thomson scattering data and despite the relatively small data set, the oscillation is clearly visible in the two-dimensional temperature evolution and the temperature profiles deviate significantly between trough and crest. Since the coherence time of the probed oscillation is much shorter than the laser repetition frequency, yet always excited within the 100 frames recorded around the laser measurement, the conditional average method could reconstruct the plasma dynamics without prior knowledge of the oscillation phase. This was the prerequisite for analyzing intermittent events, which can be regarded as fulfilled.

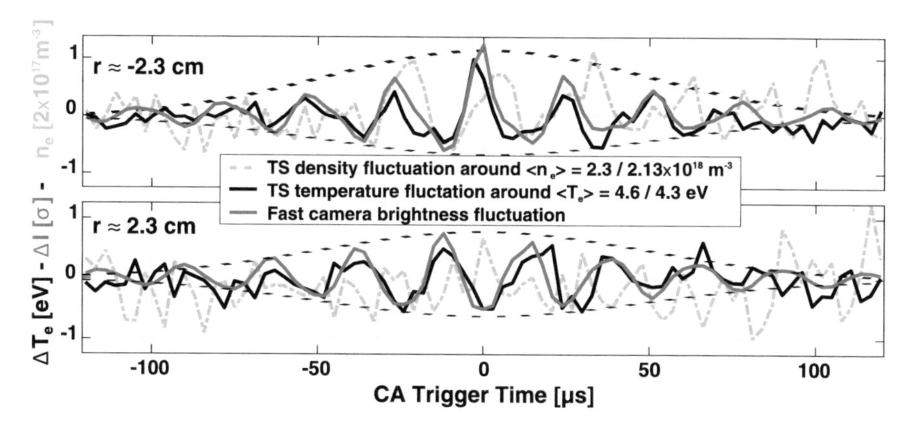

Fig. 8.18 Time traces of brightness, density and temperature in the upper and lower maximum for each quantity

Since the measurement conditions for the first and third case from Fig. 7.3 were unstable but showed similar dynamics in the brightness and temperature, a detailed presentation is omitted here. The fourth case with 400 A and 100 sccm was somewhat stable and delivered enough signal despite the overall sensitivity declines. However, the visible oscillation in this discharge setting shows several competing or coexisting oscillations, which require more specific conditional triggers and thus longer time sequences for an analysis with similar accuracy as in the 350 A 50 sccm case.

8.2.2 *Intermittent Bursts in Deuterium Discharges*

For the final analysis of intermittent bursts in the Deuterium discharge with 300 A and 55 sccm a total of 42,000 laser shots were used. The Thomson scattering subsets are created in the same way as the previously described Argon discharges, but with the conditional trigger in the far edge as shown in Fig. 8.9. The average number of laser shots contained in each subset decreases from 12% down to 5% of the total number of laser shots for a brightness threshold of $I_{\text{CA}} \geq 1.25$ to $2\,\sigma$. The spatiotemporal resolution is again reduced to 2 (pseudo 0.5) cm and 3 μs, respectively, to minimize uncertainties from low photon numbers in the fitting routine. The space resolution is sufficient according to the structure size of the intermittent events and kept constant to ensure comparable results, while the time resolution is fixed at the frame rate for the duration of the experiment. The fastest dynamics still visible at high-speed (715 kHz) are smoothed yet still distinct as the intermittent event lasts for at least a few μs ($t_{FWHM} \approx$ 3–6 μs).

Figure 8.19 shows the relatively stable signal strength over the course of the experiment and the negligible level of stray-light in the upper picture. The Raman signal declines by 30% over the whole experimental day, thus only a 5% decrease

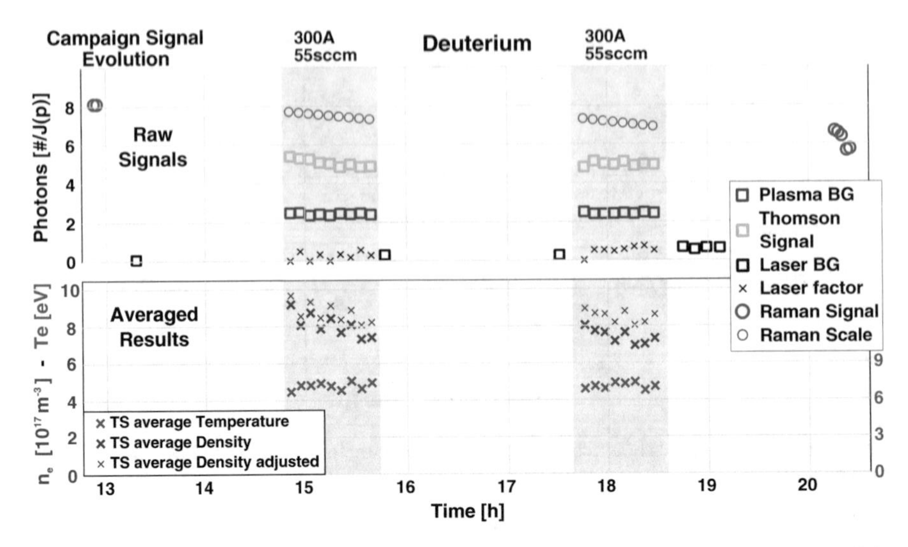

Fig. 8.19 Signal evolution during experiment for normalized Raman, Thomson and stray-light signal intensities, with the plasma background emissions scaled to indicate relative signal level in the upper half picture. The lower half shows average plasma values based on individual files with 3423 laser shots each

is attributed each during the first and second measurement sequence, since they last for only an hour each. The 3% decline of the laser energy is already factored in by the normalization.

The lower picture in Fig. 8.19 shows an almost constant mean density in the first run and a stable mean temperature throughout the discharge. For this discharge setting the plasma background emissions are the largest contributor of unwanted signal during the signal processing. This raised the question, whether the increased brightness observed with the fast camera would also be present in the plasma background of the TS measurements and thus lead to incorrect background subtraction.

To test the correlation between the plasma background collected within the range of 518–547 nm and the brightness oscillation perceived by the fast camera, the exposure time of the iCCD was set to 1 μs and triggered without laser. No radial dependence or correlation between the radially averaged signals could be established in Deuterium discharges and thus the background variations were assumed to be constant for all conditional average subsets. The lack of signal correlation could be connected to the different imaging areas and the absence of line radiation in the spectral range of the spectrometer. Since the collection cone of the plasma background light extends throughout the whole plasma and the depth-of-field is only about 5 mm, all localized structures in observing the plasma background are smeared out. Additionally, the high powered discharge conditions with low gas feed do not lead to high emissions of molecular lines. However, when using TS with CA in recombining conditions in Deuterium [7] or other gases with high photon emissions, especially

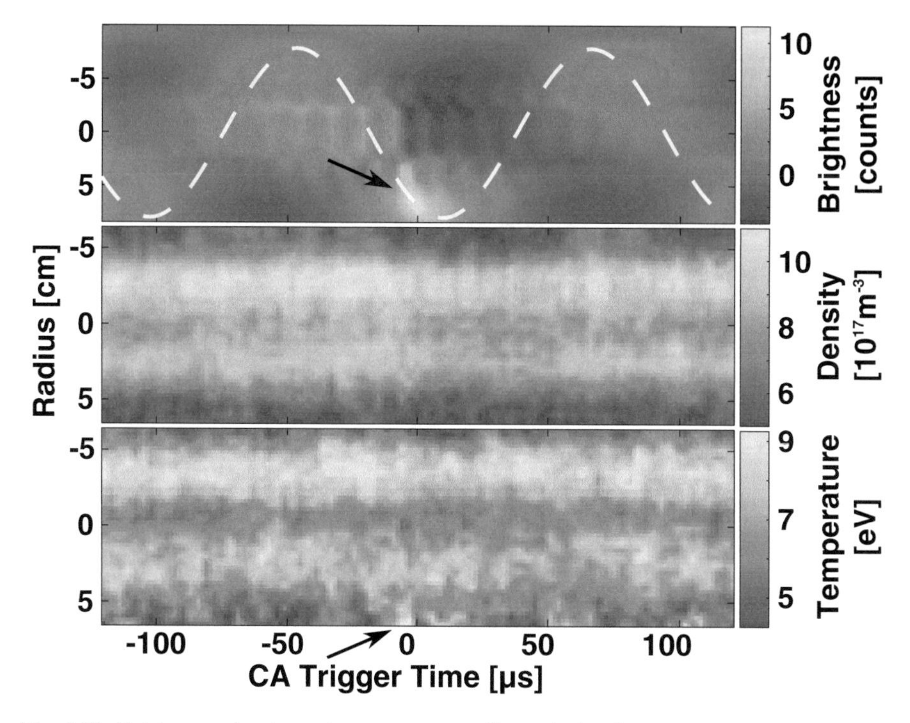

Fig. 8.20 Brightness, density and temperature profile evolution for a conditional average with $\sigma \geq 1.5$ at $r = 7.25$ cm for a Deuterium discharge

around the laser wavelength, the time dependence of the plasma background must be considered.

The conditionally averaged brightness, density and temperature profile evolutions for this Deuterium discharge are shown in Fig. 8.20. The condition of the intermittent events surpassing $1.5\,\sigma$ was set at a radial position of $\mathrm{r} = 7.25$ cm, at which the highest average brightness is visible in the upper picture. The central and lower picture are comprised of 81 Thomson scattering profiles with each approximately 7.4% of the total shots. The fast oscillation in the plasma center and the slow oscillation in the plasma edge can be identified only in the brightness, while the bulk density increases up until the ejection at $\mathrm{t} = 0\,\mu$s. Fast variations in density and temperature are not observable, but the ejection itself leads to a sharp increase in the edge temperature, which is preceded by a temperature and density peak at the opposite radial side about 15–25 µs before the ejection. For a better comparison of bulk and edge plasma dynamics Fig. 8.21 shows the averaged bulk ($\mathrm{r} \leq \pm 3$ cm,$\mathrm{r} \leq \pm 7.25$ cm) density, brightness and temperature in the upper picture, while the same quantities are presented for the edge ($\mathrm{r} = 6.5$ cm) in the lower picture. The fast camera brightness and density are shown in bold traces in the upper picture. The increase of the average density reaches 5% above average and drops to 5% below average within 30–40 µs and follows the similar but smoother brightness trace. The bulk plasma

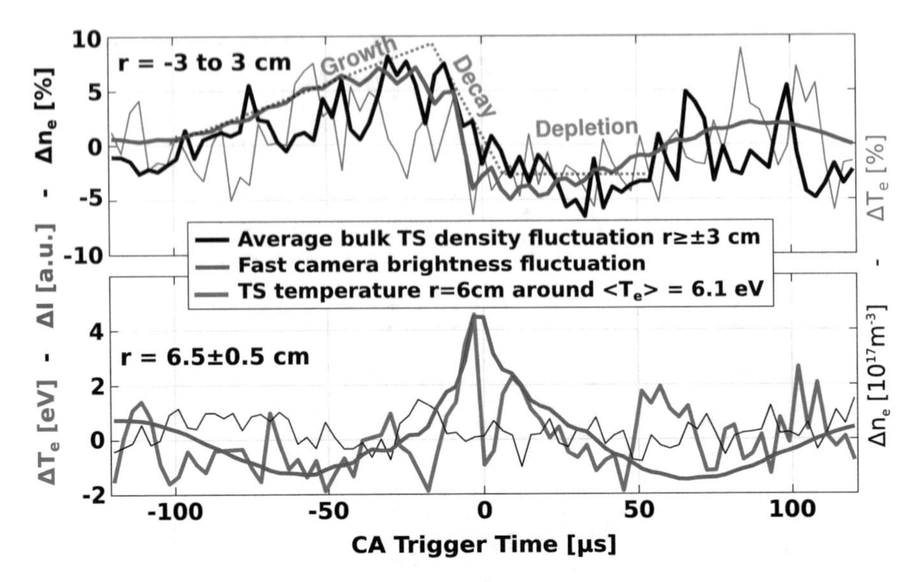

Fig. 8.21 Time traces of brightness, density and temperature in the bulk and edge plasma for each quantity

temperature trace shows a weaker resemblance with the brightness, but not opposing the trend. In contrast, the brightness evolution in the edge follows more closely to the temperature, while the density seems unrelated. Therefore, in both cases the plasma pressure compares well to the brightness, but in the bulk the density is more responsible for the brightness fluctuation and vice versa in the edge. The peak temperature reaches 10 eV with an increase of 4 eV one frame (3 μs) before the CA trigger, but synchronous to the density drop in the center. Since the maximum temperature in the bulk is also around 10 eV and the density drops by a total of 10% the expulsion of plasma around the intermittent event is confirmed. However, a density increase is not observed in the plasma edge.

The conditionally averaged measurements lack the time resolution to resolve the oscillation and the exact mechanisms behind the intermittent transport events. The slow density increase and release is only captured for one period within the 100 frames and the fast oscillations remain fuzzy. However, conclusion could be drawn from the free-running fast camera measurements for statistics and conditional averages with higher frame rates (cf. Fig. 8.9). The slow $m = 1$, 10 kHz oscillation appears in the conditional average and seems to promote the intermittence, while simultaneous plasma expulsion to both sides are present but far less likely ($\sim\frac{1}{10}$) and/or weaker. The inner modes ($m = 1$ or $m = 2$) grow rapidly before the ejection and seem directly linked to the creation of the intermittent event. A plausible explanation for the intermittence is that the slow, outer oscillation pushes the growth rate of emerging inner modes by variation in the pressure gradient. As in the standalone CA brightness, the fast oscillation roughly grows and decays within 75 and 25 μs,

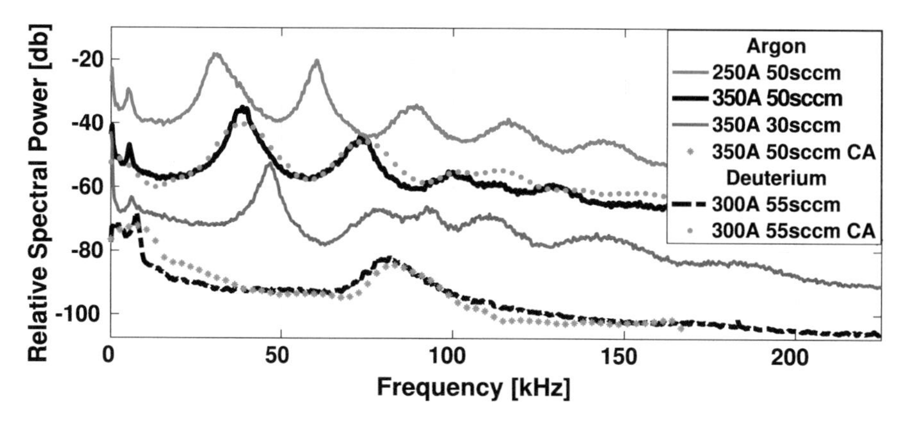

Fig. 8.22 Spectral power of free running videos and CA frames, averaged over entire radius

respectively, indicated by the dotted gray lines in Fig. 8.21. The time of the growth corresponding to roughly half the period time of the slower oscillation, while the decay coincides with the temperature peak at the edge. After the decay the bulk density is depleted for a 50 μs period until it recovers.

To consolidate the findings by conditional averaging towards the general fluctuations in the plasma and interpret the oscillations, Fig. 8.22 compares the averaged power spectra of conditionally averaged events (gray dots/stars) to the standalone recordings (solid/dashed lines). As the CA spectra are almost unchanged and thus considered to be representative for the general discharge conditions, the density and temperature dynamics associated to the mode structure in Argon and filaments in Deuterium by Thomson scattering can be generalized. Of particular interest are here the statistical fluctuation properties in Deuterium for their effect on erosion and plasma transport in the edge of the discharge.

In addition to the analyzed cases of Argon and Deuterium, further Argon power spectra are presented. For the first three Argon cases the frequency of the dominant mode rises as the pressure rises (mainly through temperature), while the higher harmonics blend together. This increase is either pressure related or an effect of the lower effective mass by an increasing fraction of double ionized Argon ions.

8.2.3 Discussion

Dissecting the mechanism behind the turbulence in PSI-2 might be more complicated than the usual linear plasma device (homogeneous field, gaussian profile) and therefore require a dedicated experiment, while previously conducted experiments of plasma dynamics on PSI-1 and PSI-2 give some additional ideas for possible dynamics [5]. A subbranch of driftwaves, called ion drift waves (IDW), has been observed in Krypton discharges, where the IDW generates mode structures ranging from $m = 1..6$ (with 5 being the strongest) by finite Larmor radius effects. The frequencies observed by Langmuir probes range from 5 to 70 kHz with a phase velocity of roughly 2 km s^{-}1. Plasma rotation velocities of ions in general have been observed in PSI-1 [8], where corresponding frequencies of 5–15 kHz were found to scale with input power, magnetic field angle in the source and the mass of the used ion species. Although the measurements were conducted in the differential pump stage or in a slightly different magnetic arrangement of PSI-1, the oscillations found in this work fall within the previously observed drift wave range, and perturbation are expected to be carried through the high-field region into the exposure chamber due to the fast axial transport.

For symmetric linear machines with a well defined shape, there are already a number of different causes reported for turbulent and non-turbulent structures, which include neutral wind in the SOL [9], flow shear [10–13] and general drift wave turbulence caused by strong pressure gradients [14] and the reduction of neutral density [15]. As the primary usage of PSI-2 is dedicated towards plasma-material interaction studies and reaching a high plasma fluence, other boundary conditions for an easier plasma characterization are subordinate, e.g. the magnetic field is not constant and slight asymmetries are tolerable. However, the flux expansion into the exposure chamber is another potential source for low-frequency instabilities, which can be triggered by a double layer (spatially isolated, rapid changes in plasma potential) [16, 17].

The pressure gradients in PSI-2 responsible for generating drift wave turbulence are in dire need of clarification. On the one hand, Langmuir probe data provide strong inner and outer gradients in the hollow profile, while the time-resolved TS measurements combined with the fast cameras show much flatter overall profiles, but strong temporal variations, whose dynamic accelerate with decreasing ion mass.

On the other hand, the fluctuation statistics and power spectra are similar for Langmuir probe and fast camera. Nevertheless, resulting from various sources of turbulence, the generation and expulsion of coherent structures is a property of many linear plasma machines. The experimental results show pressure and brightness fluctuations in a typical drift-wave (turbulence) frequency range, coherent, axially elongated oscillations or filaments expelled with radial velocities of a few km/s and a radial or azimuthal size of about one cm. Without the examination of a wider range of plasma dynamics by a systematic parameter scan within discharges, the ion mass dependence is the only parameter variation, confirming a dampening by the higher inertia of the increasing ion mass. An advanced interpretation of the dynamics is beyond the scope of this work for there are many particular destabilizing mechanisms to choose from. However, the key statistical properties, which are expected to be similar in all magnetically confined devices, could be confirmed within PSI-2 with its "unusual" magnetic configuration.

Furthermore, a feedback loop like in a predator-prey-model can be excluded as an explanation for the plasma dynamics as it would require an increased amplitude of the outer oscillation *after* the ejection, which is not observable. A feedback loop shifting energy between different modes would also cause longer coherence times, since the existing modes exchange/shift energy between each other but never fully decay.

However, one important experimental finding is the observation of temperature fluctuations related to brightness fluctuations, which are generally neglected, while a density-brightness relationship is inferred in Deuterium and Argon [18, 19]. A theoretical evaluation with Helium showed both density and temperature to play a role [20] at the frequencies observed in this work. Since in this work no spectral filter was usable for Argon, the *total* photon emission received by the camera is responding to temperature fluctuation. Each of the transition lines (cf. Fig. 4.4) is emitting photons based on temperature and collision-rate dependent population densities, hence this relationship might only be possible by the relatively high electron temperature in the first place and low-temperature Argon discharges with high density (fluctuations) could behave differently. Nonetheless, the general possibility of temperature fluctuations must be acknowledged in PSI-2, especially since they influence crucial processes in PWI and bias predictions if not considered (cf. Chap. 2).

8.3 Impact of Hot Plasma Filaments on PWI Processes

The experimental findings of intermittent filament ejection in Deuterium are now reconciled with the focus on the PWI processes, based on the impact of their statistical fluctuation properties, outlined in Sect. 2.3. With the examples in the previous section the conditional average of intermittent spatiotemporal plasma structures was analyzed by time-resolved Thomson scattering and the PDFs of the brightness fluctuations can be qualified by TS.

Although the definition of intermittence only holds in the far edge, it is clear that the plasma ejections are still related to oscillations, while non-linear interactions between the modes cause the deviation from a stable rotation. The amplitude requirement for defining filaments varies in the literature, but either a certain threshold above background pressure is given or well above the standard deviation of a fluctuating signal. The thresholds used in this work are rather on the lower end of the definition spectrum to include more shots and lower the error bars. However, to extract the properties of the filaments and the corresponding plasma conditions creating them, a higher threshold is desirable.

Figure 8.23 shows a series of profile snapshots at the ejection time with increasing CA thresholds and compares results for these triggers on either side of the plasma column. Both trigger positions are on the inner side of the pictures and marked with the dashed red line. From high to low trigger the color codes change from black to green accordingly with black in the background for better visibility, since the lower signal causes higher error bars. The left side is part of the previous analysis, where the raised temperature and density are prominent above the average profiles in gray. While the density seems to further extend the profile to the right with higher CA thresholds, the temperature peaks at similar figures at the edge. This supports the picture of plasma being ejected from the bulk or peak region of the column, because

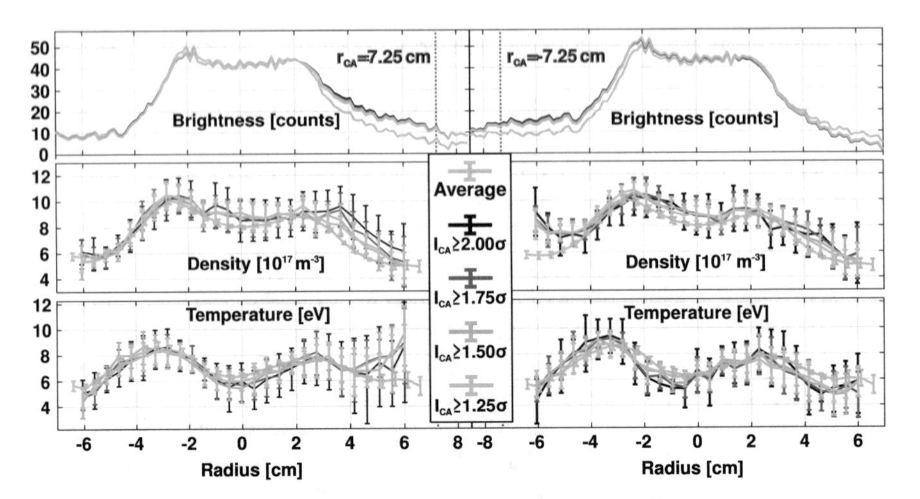

Fig. 8.23 Radial profiles of fast camera brightness,density and temperature for CA triggers at ±7.25 cm during the plasma ejection. The increasing trigger thresholds are overlayed by the average profiles of each quantity. The different radial profile extents are caused by available signal level

only a fast, adiabatic transport process like the filaments are capable of reaching temperatures corresponding to peak value of the discharge. An independent heating process seems unlikely, since the density extension suggests plasma being transported rather than locally heated. With increasing strength of the ejection, merely more plasma is ejected, causing a higher density and increased brightness.

A very different picture is seen on the right side with the CA trigger at $r = -7.25$ cm, where the increased brightness seems solely caused by an increase in density well above the error. Although the temperature profile changes are subtle, an increased gradient on both sides of the negative (radius) peak region seems to be connected to the ejections process. Asymmetries were found already in previous figures depicting CA profiles and their evolutions, as well as in the ejection probability (based on trigger threshold). Looking at the gradients and the meaning of turbulent transport a simple relation could be inferred as follows. The steeper gradient at negative radii corresponds to a lower transport coefficient and vice versa for the much weaker decay on the opposite side, which is caused by a higher number (probability) of filaments being ejected on that side. Density and temperature peaks are slightly lower as well. The reason(s) for this asymmetry could be manifold from ranging from asymmetries in the plasma source over to chamber or magnetic field related inhomogeneities.

Another important observation for the analysis of Thomson profiles towards lower signal levels is the reproducibility of the results, despite the large error bars. The fitting routine works robustly, while smooth deviations with lower signal levels would be unexpected.

The skewness profiles of brightness, pressure and temperature are shown for four CA cases in Fig. 8.24, where the saturation of the lines indicate the CA threshold

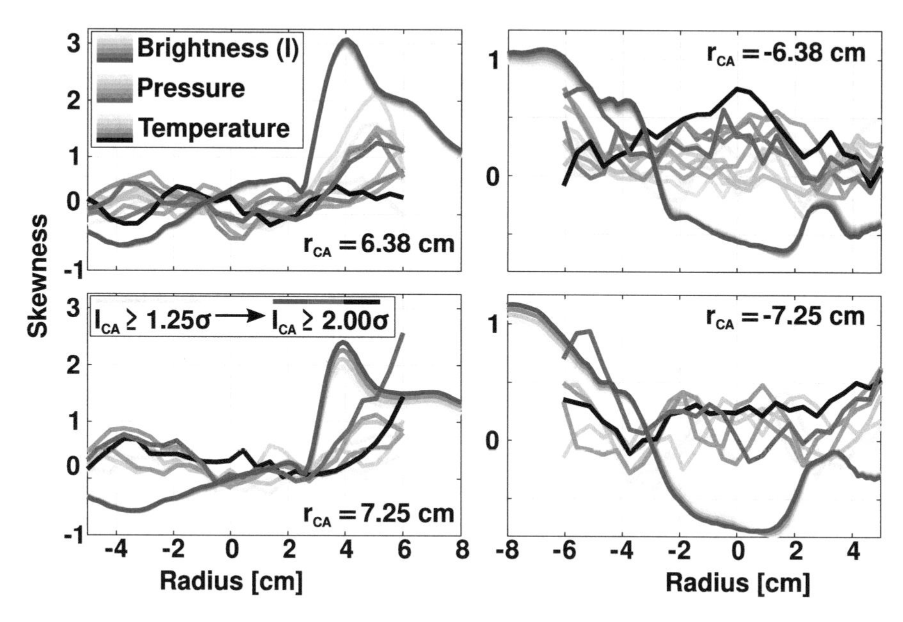

Fig. 8.24 Skewness profiles of brightness, pressure and temperature for four indicated CA trigger radii in Deuterium. The saturation of the curves increase towards higher CA amplitude thresholds. Note the threefold higher skewness scale on the left

ranging from 1.25 to 2 σ. The central skewness values fluctuate around zero or slightly positive, while the triggered regions of interest are in the middle of the figure. Here the pressure skewness is exclusively positive and usually increasing with higher σ. As observed before, the CA triggered brightness at negative radii respond less to the edge temperature, while on the left the temperature skewness is always positive at the edge, similar to the pressure. Albeit only 81 time steps of each CA time series are used for the skewness calculation, the fluctuation statistics of brightness, pressure and temperature share the same bias towards a positively skewed PDF.

An illustration of the fluctuation statistics with the corresponding shares of background and intermittent parts, representing diffusive and turbulent/convective transport is shown in Fig. 8.25 for the edge of a Deuterium discharge. The intermittent share is fitted by the PDF of a Γ-distribution with a skewness of 1.4, while the measured overall brightness trace has a skewness of 0.95, since the mean values differ.

A qualitative physical picture could explain the background fluctuations arriving through slow diffusion with equilibrating particles and thus causing only minor excursions from the average, while the fast, turbulent and adiabatic transport delivers the large amplitude events. Therefore, both shares are needed for an accurate fit and considered for demonstrating the effect of the skewness when calculating sputtering rates. This shall be done only as an example for Deuterium on Tungsten with the sputtering yield fit function [21], shown in Fig. 2.2 on the left. Using only a one-dimensional energy distribution and a sheath acceleration of $3\,k_B\,T_e$, the sputtering yield function for a Maxwellian energy distribution similar to the right side of Fig. 2.2 is obtained, which is about 1.5 times smaller, likely caused by the exclusion of the angular distribution. Since the filaments are coherent structures, adiabatically transporting plasma from hotter regions, the realizations of temperatures rather than projectile energies were tested. No additional radial velocity was assumed. Various realizations of Gaussian and Γ distributed temperatures with a length of 2×10^6 were used to compare their effective sputtering yields to the mean temperature sputtering yield. The top pictures of Fig. 8.26 show linear (left) and logarithmic (right) sputtering rate curves for mean temperature, a Gaussian distribution, a mix of mean or Gaussian and Γ distribution, and finally a pure Γ distributed temperature realization with a skewness of $S = 2$. The difference between mean temperature and Gaussian distribution are small and negligible when the Γ distribution is added. Significant increases in the effective sputtering rate are visible, while the relative increases are tremendous in the sub-threshold temperatures around 10 eV, although the overall sputtering rate is still extremely low. Towards a temperature of 100 eV, the relative increase drops to 10%, since the sputtering rate is less steep as compared to the threshold region. The threshold itself, e.g. defined at a rate of 10^{-5} drops from 20 to 30 eV down to $\sim$15 eV. In the bottom half of Fig. 8.26 the influence of the skewness is depicted. Skewness values ranging from 1 to 2 are commonly reported in the literature and also found in PSI-2, while the share of turbulent transport compared to the total transport is estimated in the range of 50% and higher. Below a temperature of 10 eV, tremendous increases of the effective sputtering yield are visible, due to the surpassing of the energy threshold. At $S \geq 2.5$ towards 100 eV the trend reverses, since the rate flattens and the realizations contain frequently values above 2 keV, which had to be capped for calculation stability.

The fluctuation statistics of the plasma conditions in the high powered Deuterium discharges used in this work are indicated with a gray background in Fig. 8.26. While determining the range of the skewness is straightforward, the estimate for the shares of Gaussian background and intermittent fluctuation parts could be based on Fig. 8.25. The areas covered with linear axis could serve as a rough estimate for the fluctuation state shares, ranging from a quarter to half the total points. Comparing the sputtering rates for this case ($S = 1$ and 50%/50% share) at the temperature of up to 10 eV in Deuterium plasmas in PSI-2, the sputtering rate increases from 2×10^{-13} to 4×10^{-10}. This three order of magnitude increase is however extremely far from any measurable erosion, even at steady state operated machines like PSI-2. Substantial sputtering rates are reached upwards of 50 eV, which are commonly simulated by a bias towards samples surfaces. At these temperatures, the sputtering rate enhancement decreases to less than a magnitude, at which a non-Gaussian temperature fluctuation on top of the bias could even fall below the uncertainty of the employed diagnostic accuracies.

When determining the lifetime of the wall, the sputtering yield translates to erosion in thickness with a given plasma density and exposure time. In this regard, a certain sputtering yield must not be exceeded to remain below a "threshold", e.g. 10^{-7} to 10^{-5}, at which the erosion is below the thickness of the first wall layer (e.g. 2 mm

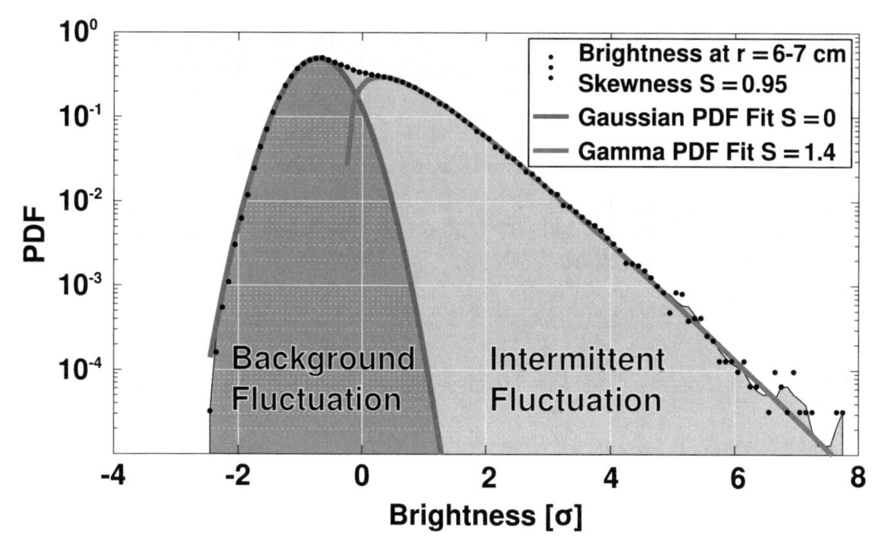

Fig. 8.25 Illustration of shares for background and intermittent fluctuations, approximated by Gaussian and gamma PDF at the edge of a Deuterium discharge

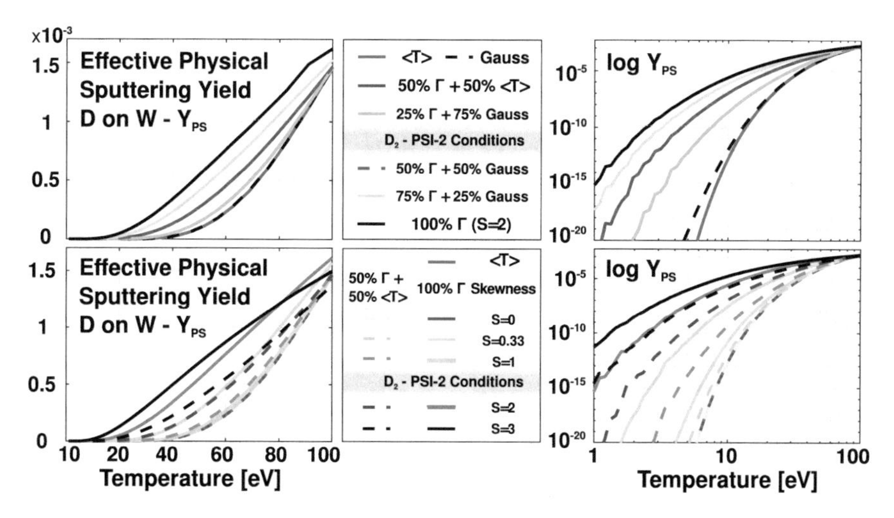

Fig. 8.26 Influence of energy distribution on effective physical sputtering yields on linear (left) and logarithmic (right) scale. Sputter rates are compared for different composition (top) and depending on skewness (bottom)

tungsten). Although this sub-threshold transition is blurred by turbulent fluctuation, the temperature at which the sputtering yield is below 10^{-5} can be determined for the different fluctuation states of the plasma edge. A horizontal line in Fig. 8.26 at 10^{-5} crosses the mean temperature line at 35 eV, while for the highest degree of turbulence the line is crossed at only 10 eV. Therefore, the determination of the acceptable sputtering yield, corresponding to a particular layer thickness and plasma density, must include additionally the temperature fluctuation statistics to choose the correct maximum temperature.

References

1. Ohno N, Furuta K, Takamura S (2004) Visualization of intermittent blobby plasma transport in attached and detached plasmas of the NAGDIS-II. J Plasma Fusion Res 80(4):275–276
2. Antar GY, Yu JH, Tynan G (2007) The origin of convective structures in the scrape-off layer of linear magnetic fusion devices investigated by fast imaging. Phys Plasmas 14(2):022301. (article id. 022301, 10 pp)
3. Windisch T, Grulke O, Naulin V, Klinger T (2011) Intermittent transport events in a cylindrical plasma device: experiment and simulation. Plasma Phys Control Fusion 53(8):085001. (article id. 085001, 18 pp)
4. Reiser D, Ohno N, Tanaka H, Vela L (2014) A plasma source driven predator-prey like mechanism as a potential cause of spiraling intermittencies in linear plasma devices. Phys Plasmas 21(3):032302. (article id.032302)
5. Klose S, Bohmeyer W, Laux M et al (2001) Investigation of ion drift waves in the psi-2 using langmuir-probes. Contrib Plasma Phys 41(5):467–472
6. D'Ippolito DA, Myra JR, Zweben SJ (2011) Convective transport by intermittent blob-filaments: comparison of theory and experiment. Phys Plasmas 18(6):060501. (article id. 060501, 48 pp)
7. Pospieszczyk A, Reinhart M, Unterberg B et al (2013) Spectroscopic characterisation of the psi-2 plasma in the ionising and recombining state. J Nucl Mater 438:S1249–S1252
8. Meyer H, Klose S, Pasch E, Fussmann G (2000) Plasma rotation in a plasma generator. Phys Rev E (Stat Phys Plasmas Fluids Relat Interdiscip Top) 61(4):4347–4356
9. Krasheninnikov SI, Smolyakov AI (2003) On neutral wind and blob motion in linear devices. Phys Plasmas 10(7):3020–3021
10. Carter TA (2006) Intermittent turbulence and turbulent structures in a linear magnetized plasma. Phys Plasmas 13(1):010701. (article id. 010701, 4pp)
11. Windisch T, Grulke O, Klinger T (2006) Radial propagation of structures in drift wave turbulence. Phys Plasmas 13(12):122303. (article id. 122303, 7 pp)
12. Windisch T, Grulke O, Naulin V, Klinger T (2011) Formation of turbulent structures and the link to fluctuation driven sheared flows. Plasma Phys Control Fusion 53(12):124036. (article id. 124036, 9 pp)
13. Manz P, Xu M, Müller SH et al (2011) Plasma blob generation due to cooperative elliptic instability. Phys Rev Lett 107(19):195004. (article id. 195004)
14. Thakur SC, Brandt C, Cui L et al (2014) Multi-instability plasma dynamics during the route to fully developed turbulence in a helicon plasma. Plasma Sources Sci Technol 23(4):044006. (article id. 044006)
15. Saitou Y, Yonesu A, Shinohara S et al (2007) Reduction effect of neutral density on the excitation of turbulent drift waves in a linear magnetized plasma with flow. Phys Plasmas 14(7):072301. (article id. 072301, 7pp)

16. Aanesland A, Lieberman MA, Charles C, Boswell RW (2006) Experiments and theory of an upstream ionization instability excited by an accelerated electron beam through a current-free double layer. Phys Plasmas 13(12):122101. (article id. 122101, 10 pp)
17. Carr J, Galante ME, Magee RM et al (2013) Instability limits for spontaneous double layer formation. Phys Plasmas 20(11):113506. (article id. 113506, 7 pp)
18. Banerjee S, Zushi H, Nishino N et al (2012) Fast visible imaging and edge turbulence analysis in quest. Rev Sci Instrum 83(10):10E524
19. Oldenbürger S, Brandt C, Brochard F, Lemoine N, Bonhomme G (2010) Spectroscopic interpretation and velocimetry analysis of fluctuations in a cylindrical plasma recorded by a fast camera. Rev Sci Instrum 81(6):063505–063505-7
20. Ma S, Howard J, Thapar N (2011) Relations between light emission and electron density and temperature fluctuations in a helium plasma. Phys Plasmas 18(8):083301. (article id. 083301, 14 pp)
21. Eckstein W (2007) Sputtering yields: experiments and computer calculations from threshold to MeV energies. Springer Berlin Heidelberg, Berlin, Heidelberg, pp 33–187

Chapter 9
Summary and Conclusion

The aim of this work was to investigate the impact of plasma dynamics on PWI processes with a newly developed time-resolved Thomson scattering diagnostic, which is motivated by the potential influence of plasma fluctuations on PWI. In the course of this work the laser based plasma diagnostic setup was designed, built and installed on PSI-2. Accompanied by Langmuir probes and two fast imaging cameras, a photon counting algorithm and conditional averaging, the Thomson scattering system proved the ability of time-resolved measurements. Two selected discharge cases were analyzed with a time resolution of 3 μs as a proof-of-principle for this analysis. The conclusions for this work include the implications concerning the Thomson scattering itself and other diagnostic directly or indirectly involved, especially with respect to the diagnostic limits to detect filaments in plasma turbulence. These findings are then applied to PWI in general as they are of interest for erosion in fusion reactors and to the operation of the linear plasma generators PSI-2 and JULE-PSI.

The new laser based plasma *Thomson scattering diagnostic* presented in this thesis contributes the capability of measuring density, temperature and velocity profiles with a radial resolution of up to 3 mm in PSI-2. The lower bound of the electron density error is defined by the Raman scattering cross-section of 8%, while a statistical error of lower than 10% is generally achieved within 5 m for low stray-light levels and higher densities, while poor conditions and low density discharges require up to one hour of integration time. With lower spatial resolution, these limits shift to shorter times accordingly. The temperature boundaries for TS are 1 and 20 eV, originating from the the notch filter blocking the central spectrum and the width of the iCCD, respectively. The requirements for a 10% statistical error are about twice the density requirements and usually dominate the instrumental lower bound caused by the spectral convolution.

The results obtained for all discharges in Deuterium, Helium, Neon and Argon show a reasonably good agreement for TS density and temperature profiles compared to Langmuir probe measurements, considering the combined errors of both methods.

M. Hubeny, *The Dynamics of Electrons in Linear Plasma Devices and Its Impact on Plasma Surface Interaction*, Springer Theses,
https://doi.org/10.1007/978-3-030-12536-3_9

The generally hollow profiles and peak positions of density and temperature are mostly confirmed, while the drop of both parameters towards the plasma center is much weaker in the Thomson profiles. The discrepancies of the plasma parameters are thought to be caused by a combination of diagnostic uncertainties and different locations, while the absolute figure deviations are well within a factor of 1.5 to 2.

The ability to preserve temporal information with Thomson scattering was achieved by developing a *photon counting algorithm* for an iCCD camera. The algorithm allows flexible measurement durations without sacrifice of signal intensity, noise level or spatial resolution. Then, the synchronization of laser and visual imaging transfered the time resolution of the fast camera to the Thomson scattering system by the knowledge of the plasma state around the laser shot. The integration of shots to a Thomson spectrum is selected by the *conditional averaging method* applied to the fast camera data.

A successful test of this procedure was achieved for a high power, low gas-feed Argon discharge. The discharge with a dominant $m = 1$ mode was analyzed by the time-resolved Thomson scattering method and the cause of the 40 kHz brightness oscillation was identified as a temperature fluctuation with an amplitude of up to 1 eV and a total excursion of about 1.5 eV. The fluctuation pattern extracted by the conditional average is consistent with free-running recordings of the plasma fluctuations and hence representative. Therefore, the common assumption of an absence of temperature fluctuations does not hold in this case and should be tested systemically for other discharge cases. The mean brightness in Argon discharges is mostly generated by the rotating structures, leading to serious implications for other diagnostics. In the most extreme case all of the light collected by spectroscopy stems from a radially localized structure rotating at 10–50 kHz, since the temperature excursion amplifies the photon emission drastically. The integration time for an I-V-curve of the Langmuir probe is with 20 ms much longer than the oscillations. Especially averaging in the exponential region of the characteristic leads to interpretation problems.

The structural properties and dynamics of the turbulence were analyzed with fast framing cameras with up to 1.3 MHz. Ion saturation current measurement in selected discharges showed consistent spectral features compared to the fast camera measurements. Plasma discharges with Argon and Neon exhibit *oscillating multi-mode structures with coherence times in the order of* 50–100 μs, while for Helium and Deuterium an additional slow oscillation appears at the plasma edge. The structural and spectral properties are attributed primarily to driftwave turbulence, while other causes can not be ruled out completely. With lower atomic mass some of the plasma dynamics in Helium and Deuterium accelerate beyond the time resolution of the fast cameras, likely because the inertia of the ions is less effective in slowing down the dynamics. Additionally, all discharges were carried out with the lowest gas feed possible with each gas type minimizing ion-neutral collisions, which would otherwise stabilize instabilities by friction. The plasma diameter defined by steady state pressure profiles and average brightness is increasing from Argon to Neon and Helium, while towards the Deuterium discharge condition a threshold seemed to be surpassed and the increased transport is lowering pressure and plasma size again.

In high power Deuterium discharges the SOL of PSI-2 exhibits *large scale brightness fluctuations*, which are identified as *plasma filaments causing turbulent transport* far beyond the plasma bulk. With radial velocity estimates ranging up to $6\,\mathrm{kms}^{-1}$, these (axially) coherent structures are similar to filaments found in the SOL of fusion devices. Moreover, statistical properties like a positively skewed PDF of fast camera brightness and ion saturation current and intermittency were confirmed. Therefore, the final goal of this work was to qualify the time resolution and accuracy of the Thomson scattering setup for investigating the filaments in Deuterium discharges in PSI-2. While the temporal resolution of the new diagnostic did not suffice to clarify the generation process in the core plasma, essential properties of filaments were retrieved beyond the error margin of the method. A density built-up and release of 10% of the bulk (-3 to $3\,\mathrm{cm}$) plasma was discovered prior to the ejection of plasma. In contrast, the brightness increase in the far edge of the discharge corresponds primarily to a temperature increase on the more active side of the discharge. The temperature and pressure evolution in temporal vicinity (conditional average) of the ejection exhibit a positively skewed PDF, which aligns the statistical findings of Langmuir probe, fast camera and now Thomson scattering. Consequently, the implications of such a skewed PDF are concluded in the following.

Besides controlling the burning plasma in the center of a fusion reactor, exhaust and transition to the solid walls and structure of a plant are the crucial factor for a feasible fusion power plant operation. Although some degree of wall erosion is practically inevitable, the lifetime of the wall components dictates the duty cycle of a reactor and hence all known sources of enhanced erosion, disruptions and ELMs, must be actively mitigated during reactor operation. Intermittent filamentary transport is a universal feature of magnetically confined plasmas and can not be suppressed to this date.

Modeling or simulating the effects of (micro-)turbulence on the wall erosion requires similar time scales (months/years) or extremely precise measurements, as the erosion margins are only a few mm. The filament properties found during the investigation of the Deuterium discharges emphasize the importance of time-resolved measurements of edge plasma properties, since the brightness fluctuations correspond mainly to temperature fluctuations, whose mean values obscure the influence of skewed PDF on nonlinearly temperature dependent processes like sputtering.

A fluctuation distribution with a skewness around or above unity is common in the literature, however the association to density, temperature or pressure is difficult. The effective sputtering yield based on non-Gaussian temperature fluctuations virtually eliminates the "sub-threshold" operational point and leads to strongly increased sputtering yields up to a mean temperature of $100\,\mathrm{eV}$.

Furthermore, the surface area of the first wall adsorbs and recycles fuel, ash and impurities of the reactor and these processes are highly sensitive to density AND temperature fluctuation. Detailed knowledge of the fluctuation statistics is thus necessary for meaningful predictions and possible strategies to suppress or mitigate adverse effects of the plasma filaments.

The application of Thomson scattering on PSI-2 provides useful feedback to its possible application and limitations on JULE-PSI, while the measurements themselves are important for the utilization of the plasma column, since a similar plasma source will be used. PSI-2 provides a realistic environment for achieving plasma conditions similar to inside of the reactor in terms of heat load, particle load and material temperature. Achieving high fluence on samples with a known particle energy distribution is the basis of the analysis and modeling of PWI processes. Spectroscopic measurements supporting Langmuir probe profiles served as sufficient consistency of discharge characterization, while Thomson scattering is generally considered as a more accurate diagnostic. However, regarding the parameter fluctuations in high-power discharges investigated in this work and the discrepancy between Langmuir probe and TS profiles, a certain ambiguity has surfaced systematically throughout most measurements. The Abel-inversion of line integrated spectroscopy requires an input profile shape, assumptions about azimuthal symmetry and ignores time-dependence, hence no clarification can be expected. Based on the presented TS measurements, the density profiles show a far less pronounced drop towards the center, thus the exclusion of this exposure area should be reviewed. Furthermore, the temperature profile plays a minor role for the particle energy deposited on the surface, while the probe bias and the plasma potential are the main contributors. Increasing the throughput of exposed samples per session and thereby decreasing the number of sample changes accelerates experiments and increases efficiency in PSI-2 and is even more important in a remote-handled device like JULE-PSI. Nevertheless, since the presented TS setup is aiming at time resolution, which is out of focus for JULE-PSI, where a more rigid, robust installation of TS would be required. The major disadvantage of TS is the high operational effort, especially related to the optical arrangement. However, if signal and laser transmission and calibration are ensured, no contamination as in the case of intrusive diagnostics would occur, while the optical signal acquisition could be used for other spectroscopic methods as well.

9.1 Outlook and Future Directions

The newly installed Thomson scattering system with the ability to measure intermittent plasma fluctuations time resolved by conditional averaging is the first of its kind and achieved a proof-of-principle status during the course of this thesis. The optics and imaging system of the cameras (CMOS and iCCD) and spectrometer are state-of-the-art, highly optimized and efficient, while the implementation and measurement involved many manual procedures and massive amounts of RAW data ($\leq$400 GB/day) due to the pilot character of the experiment. During the course of the campaigns many improvements were implemented on the fly or after analysis, while other long-term improvements could not be implemented, but mentioned as future enhancements with their potential influence on the experimental capabilities.

The benefits of an improved Thomson scattering setup are versatile, since many aspects are either improved or more precision can be achieved with dedicated configuration changes. Fundamentally necessary is a general increase of signal intensity (and SNR) by increasing laser power, reducing stray-light and background and increase signal stability. Combined, these effects could allow an order of magnitude increase in signal to be used for spectral or spatial accuracy gains. Especially an increase in spectral resolution is greatly anticipated for it allows conclusions based on the velocity profile and the deviation from a Maxwellian EEDF.

Additionally, the advanced use of the fast camera synchronization with better lenses and two cameras allows a wider range of conditions for the CA method, while more structural information will be added as well. Technically, the alignment of temporal resolutions in free-running and CA measurements with TS is possible, but so far restricted based on the available signal strength. Without this restriction, more details of the fast Deuterium oscillations are resolvable by TS.

The experimental stability during the course of a day is critical for prolonging the acquisition time. Preparations are already in place to counteract two of the most pressing issues, the decreasing window transmission by metal coatings and the beam position control. The first problem is a general concern for spectroscopy at PSI-2 and thus a rotating shutter was developed. Furthermore, the slight transparency of the needle stack beam dump is a unique feature for the position control and live monitoring of laser power, which needs to be exploited.

Considering all the above mentioned improvements and some advances in the analysis routines, a dedicated effort to maintain high signal power of the course of a full day seems within reason. The versatile operation of PSI-2 with several gases and mixtures over a wide range of power and gas-flow parameters, variable plasma termination and biasing, give rise to a number of worthwhile and interesting future investigations of fluctuations and plasma wall interactions. Regarding the goal of this work of establishing a diagnostic for accurate and time-resolved measurements of plasma filaments, the ultimate ambition is to understand the generation process and dynamics within these filaments. In conclusion of the mentioned improvements in the outlook, the diagnostic setup would allow investigating the temporal evolution of the EEDF in turbulent structures and hence give tremendous insights into the generation processes.

Curriculum Vitae

Michael Hubeny
Forschungszentrum Jülich GmbH
Institute of Energy and Climate Research
Plasma Physics
52425 Jülich, Germany
Phone: +49 2461 61 5440
Email: m.hubeny@fz-juelich.de
Homepage: IEK 4 - Plasma Physics

Personal

German Citizen, born on Mai 9th, 1986 in Wolfen, Germany
Address: Ludwig-Jahn-Strasse 22, 50226 Frechen-Königsdorf, Germany

Education

Matura (Abitur), Anne-Frank Gymnasium, Sandersdorf, Germany, 2005
B.Sc. Technical Physics, Ilmenau University of Technology, Germany, 2008
Completed 1st year course work of M.Sc. Technical Physics at Ilmenau University of Technology, 2009

M. Hubeny, *The Dynamics of Electrons in Linear Plasma Devices and Its Impact on Plasma Surface Interaction*, Springer Theses,
https://doi.org/10.1007/978-3-030-12536-3

M.Sc. Physics and Engineering Physics, University of Saskatchewan, Canada, 2012
Ph.D. (Dr.rer.nat) in Physics, Ruhr-University Bochum, Germany, 2017

Work and Research Experience

Three months internship at IPP Greifswald, Construction and Assembly, Vacuum division, 2008
External Bachelor Thesis at linear plasma device VINETA, IPP Greifswald, 2008
Teaching Assistant for 1st and 2nd year students at University of Saskatchewan, 2009–2012
Since 2017 Post Doc at the IEK-4 Forschungszentrum Jülich GmbH, working on the divertor manipulator project for Wendelstein 7-X

Printed in the United States
By Bookmasters